First published 2022
by CRC Press/Balkema
Schipholweg 107C, 2316 XC Leiden, The Netherlands
e-mail: enquiries@taylorandfrancis.com
www.routledge.com – www.taylorandfrancis.com

CRC Press/Balkema is an imprint of the Taylor & Francis Group, an informa business

Library of Congress Cataloging-in-Publication Data
A catalog record has been requested for this book

ISBN: 978-1-032-02896-5 (hbk)
ISBN: 978-1-032-02901-6 (pbk)
ISBN: 978-1-003-18571-0 (ebk)

DOI: 10.1201/9781003185710

Typeset in Times New Roman
by codeMantra

Contents

Preface

It was in mid-2018 when I was rigorously looking for reference books covering several fields of big data implementation. I stumbled upon a particular field, petroleum upstream, mid-stream and downstream, for which I could not find anything. I realized that such a book was not available, at least not one covering all the streams. Having received the opportunity to author a book on this topic, I was more than delighted to accept it and spread knowledge in the ever-growing domain of big data.

Numerous books, research papers and articles have been published in recent decades that focus on a specific stream and offer a comprehensive perspective on it. Yet, a book that sums up all the streams and acts as a one-stop shop for upstream, mid-stream and downstream in the petroleum sector is not currently available. This book, as such, may serve as a guide for anyone interested in gaining knowledge about the amalgamation of all petroleum streams and how they relate to one another. Efforts have been made to analyse and integrate the working, analogy and contribution to the complete process of generating petroleum products.

Particular attention has been paid to emphasizing the importance and inter-relation of upstream (in exploration part), mid-stream (in transportation part) and downstream (in manufacturing part), with respect to each other. A systematic approach and working of the production phase (upstream) with all its essential processes and methodology in exploration are put forward.

A number of facets pertaining to the transportation phase are discussed with the role of shipping, pipelines and LNG terminals. The importance of mid-stream (transportation phase) for connecting upstream and downstream is also stressed. This is followed by thorough argumentation on the manufacturing phase (downstream) along with its requisites consisting of refinery processes, petrochemical plants and fabrication of polymers & plastics.

Various types of methods, pertaining to a particular stream, have been described to enable readers to conclude favourable models to undertake in practical usage. The business perspective on the methodologies and their feasibility has also been discussed. Furthermore, the role of employed professionals like geologists, geophysicists, service rig operators, engineering firms, scientists, and seismic and drilling contractors has been touched upon.

To make this book a ready reference for those who are engaged in LNG whether out of an academic interest or as a professional in the field, specific sections have been devoted to explaining various related terminologies along with streams involving extraction, production, transportation, manufacturing and storage services.

Further, to ensure this book serves as a practical reference for day-to-day use, all the terms have been adopted in their prevalent unit to give a homogenized view. That being said, some terms may require adaptation to specific units for better understanding, and hence the homogenized adaptation has not been forced. The information contained in this book has been updated up to 2021.

I am confident that with the development in the big data and the petroleum sector, my efforts will prove helpful to all those concerned with petroleum streams and will motivate research scholars to further pursue the field.

About the Authors

Mr. Jay Gohil is pursuing Bachelor of Technology in Information and Communication Technology at Pandit Deendayal Energy University. He has authored three research papers, two conference papers, two book chapters and a book (this) during his academic study. His research interests include Big Data, Data Science, Machine Learning, Deep Learning, Data Mining and Artificial Intelligence, and he has communicated research work in esteemed journals in these areas. He has been a research intern at ISRO (Indian Space Research Organization, Ahmedabad, India), Ryerson University (Canada) and Shuffle (Norway) in the field of Data Science, Machine Learning and Artificial Intelligence. He is also a Google DSC Lead, Microsoft Learn Student Ambassador, Intel Student Ambassador for IoT, IBM Z Leader, AWS Community Builder and deeplearning.ai Event Ambassador.

Dr. Manan Shah has a B.E. in Chemical Engineering from LD College of Engineering and an M.Tech. in Petroleum Engineering from School of Petroleum Technology, PDPU. He has completed his Ph.D. in the area of exploration and exploitation of Geothermal Energy in the state of Gujarat. He is currently Assistant Professor in the Department of Chemical Engineering, School of Technology (SOT), PDPU, and Research Scientist in Centre of Excellence for Geothermal energy (CEGE). One of his areas of research is power generation from low enthalpy geothermal reservoirs using Organic Rankine Cycle. He was also involved in the designing of a Geothermal Space Heating and Cooling system at Dholera and doing research on hybrid setup in the renewable energy sector. Dr. Shah has received the Young Scientist Award from the Science and Engineering Research Board (SERB). He has published several articles in reputed international journals in the areas of renewable energy, petroleum engineering, water quality and chemical engineering. He serves as an active reviewer for several Springer and Elsevier international journals.

Chapter 1

Introduction

1.1 OIL AND GAS INDUSTRY

Also known as the petroleum industry, Oil and Gas industry is one of the biggest sectors or portions of all industries in the world in terms of monetary value (as high as $2 trillion per annum in 2021 [1]). Due to its inherent nature of providing the most crucial economic framework – Oil to the world, Oil and Gas industry stands apart as one of the most significant, widespread and impactful industries around the globe, especially for countries including the United States, Russia, Canada, Saudi Arabia and China. The reach of the industry is unfathomable, as it employs hundreds of thousands (directly) and millions (indirectly) of people internationally, generates trillions of dollars of revenue and serves as the backbone of any nation's economy (due to its significant contribution to gross domain product).

The industry today quenches 57% of global energy consumption while roughly accounting for the same percentage of global CO_2 emissions. It also accounts for 1/10th of total global stock market capital, 15% of global exports and 25% of Organization of the Petroleum Exporting Countries' (OPEC) gross domestic product. The industry also requires intense capital, expensive equipment and highly skilled labour [2] and supports a huge workforce of skilled and specialized professionals (10 million in the United States alone). These stats can be comprehended in Figure 1.1.

1.1.1 Influence

The influence of the industry echoes far beyond its territory, into diverse fields including scientific research, technology, politics, warfare and environmental science. While influencing these fields due to its significance, the industry itself has been impacted by several industries, of which the residual residues can still be seen today. For instance, the historical importance of Oil and Gas industry, along with its political reach, can be seen in the way the companies operate today. In the US, the repercussions of the dismantling of Standard Oil into different entities in 1911 [3] can still be seen in the current work structure. The reach and international directions of European O&G companies also portray the exploitation of resources by the European powers on territories that were earlier colonies or political subsidiaries of Europe. The relatively newer companies that emerged due to the fall of the USSR and privatization of China also cater to huge markets. Moreover, the subsequent National Oil Companies (NOCs)

DOI: 10.1201/9781003185710-1

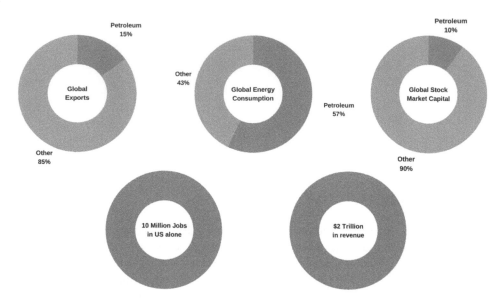

Figure 1.1 Petroleum industry statistics.

that emerged as a result of the transfer of power from the US, USSR and Europe back to the major oil-exporting nations in the mid to late 20th century also play a key role in today's supply chain [4].

Besides political impacts, one can also look at the technological advances brought through the field into the world. It began in the 1980s when mainstream oil and gas companies started utilizing digital advances through technology to make tremendous leaps into the future through various methods, including accurate measurement of reserves for understanding the level of production potential, streamline various processes in production to cut costs and make them efficient, save a great amount of time and improve the station operations around the world.

However, the extent to which technology has been integrated into the process for commendable outcomes is a mere fraction of its full potential. A single oil rig can generate millions of megabytes of data in a single day, which on proper implemented usage can prove to be extremely beneficial for decision-making process by the companies. This echoes the fact that digitization and integration of technology in Oil and Gas industry are still in its infancy with potential for several applications (as portrayed in Figure 1.2 [5]) and is something that this book explores throughout upcoming chapters.

Along with the integration of technology, the concept of automation has also been on the rise since the last decade due to various benefits that it bestows on the industry. Some of them have been summarized in Table 1.1, where automation proves to benefit drilling operations (by process speedup and reduction in safety risks), diagnosis and detection processes (by constant overlook on operations through video and data with the help of remote sensors), metering system (by automated change in configurations through sensors and programmed functions) and weather monitoring system (by automated prediction of hazards pertaining to weather for reduced safety risk and prevention mechanism).

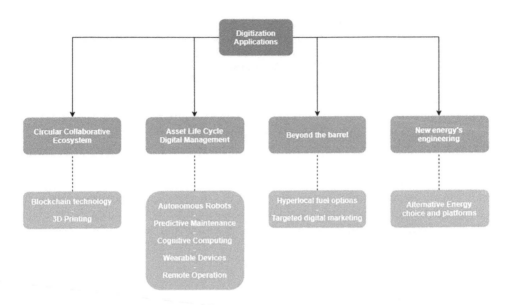

Figure 1.2 Digitization applications in oil and gas industry.

Table 1.1 Automation applications and benefits in the petroleum industry

Operation	Application and benefit
Drilling	Complete process speed-up & safety risk reduction
Remote plant monitoring	Remote access to plants and its configurations with IoT devices, resulting in reduced personnel hazards
Weather monitoring	Geological hazard activities and natural disasters prediction using autonomous sensors for actionable preventive measures and reduced construction risks
Diagnosis and detection	Real-time transmission using remote control equipment for video and data
Metering system	Real-time data return using autonomous sensors without personnel requirement along with automatic configuration adjustments

As mentioned here, Oil and Gas industry has tremendously driven various innovations in the technical field, and the associated companies have taken major steps to inculcate them into the supply chain. These innovations and steps are aimed at counteracting the lack of skilled force, slowed decreasing demand and hurdles due to governmental restrictions; while increasing production efficiency, decreasing in production time and reducing safety risks.

Furthermore, Oil and Gas industry has significant impact and reach on a key aspect that, which is the world of politics and state affairs. Due to the fact that 86% of the global oil reserves fall under the jurisdiction of the states or state companies, i.e., a significant portion of global supply falls under social control which the states have fundamental right to control, the involvement of state affairs is a no brainer for sure. However, the extent of involvement of politics and the industry is beyond our comprehension and can be slightly understood through the hereby mentioned facts about geopolitical and state affairs around the world.

The shift of oil trade from traditional regions to Asia has resulted in reduction of political influence over the Asian nations by traditional monopolies (except the continued dependency of security from United States for trade security). The western countries though will remain self-sufficient for a few decades until the reserves completely deplete. Due to the dependency of nations on petroleum for energy generation and consumption, the said geopolitical aspects about nations' Oil and Gas supremacy play a significant role in state affairs around the globe. The gas trade is no exception in the involvement of trade affairs, as the global supply chain (such as Russian gas to Europe and potential upcoming diverse Asian supplies from the Indian Ocean and Australasia) does have political influence on various decisions and deals that take place between two (or more) nations.

Therefore, as perceived till now, the scope of Oil and Gas Industry's reach is far reaching, stretching from technological advances to global state affairs. Such a level of influence on the global landscape is truly something that very few industries have been able to accomplish.

1.1.2   Challenges

It turns out that Oil and Gas industry is the only industry in the world that has consistently "lost" efficiency in its working throughout the last 100 years, contrary to the standard trend of increase in efficiency in every other industry over time [6]. This exposes a major problem with the way Oil and Gas companies operate, which is to cut costs, reduce time and maximize net production output in order to increase net profits to themselves. This mindset has cost a great deal to a smooth transition and rapid integration of cleaner & better technologies into the industry. Newer technologies are being introduced into the industry and will eventually replace the traditional working methods, but that does not appear to happen anytime soon.

There is also a shortage of skilled labour in the industry that significantly offsets the swift transitions to newer technology, change in production chain models and integration of modern equipment in the supply chain [7]. Along with this, the investors have also been in turmoil as the returns in the oil and gas industry have significantly underperformed other industries [5], leaving the viability of their investments in question. The transition of major global superpowers from tradition to contemporary strategic focus also poses a threat on oil and gas industry due to it not being the centre of attention as portrayed in Table 1.2.

According to the International Renewable Energy Agency, more than half of new utility-scale renewable power generation for electricity was cheaper than new oil and natural gas generated electricity in 2019 alone (with the former's pace not stopping anytime soon), which will add to the atrocities faced by oil and gas industry [8].

1.1.3   Future scope

In terms of persistency, the industry has recovered from major setbacks, collapses and mishaps, including Mumbai High North (BHN) Platform explosion (2005) [19], Usumacinta Disaster (2013) [20], price collapse of 2013 and recent Galveston Bay (2014)

Table 1.2 Major countries' strategic focus

Country	Government strategy	Strategy focus
United States	"Advanced manufacturing partnership" [9] "Framework for revitalizing American manufacturing" [10] "A national strategic plan for advanced manufacturing" [11] "Strategy for American leadership in advanced manufacturing" [12]	To use the internet for activating traditional industries while maintaining manufacturing industry's long-term competitiveness (Re-industrialization)
China	"Made in China 2025" [13] "Plan of action on German-Chinese cooperation on joint innovation" [14]	To create a new generation of the IT, biomedicine, bio-manufacturing, high-end equipment-manufacturing and new energy industry
United Kingdom	"The future of manufacturing: a new era of opportunity and challenge for the UK" [15]	To integrate ICT, new materials and other technologies into products & production networks, design, manufacture, supply and use of products
Germany	"Industry 4.0" [16] "High-Tech Strategy 2020 for Germany" [17]	To become a leading market and major supplier of latest generation industrial production technology
Japan	"White Paper on manufacturing industries" [18]	To boost intelligent production-line and 3D-modelling technology

and Deepwater Horizon Oil Spill (2010) [21]; that truly portrays the level of resistance to catastrophic failure that the industry possesses. The political influence in the industry can also be seen here by the rate with which recovery occurs in case of mishaps; for instance, the production restraint agreement between members of the OPEC and ten non-OPEC partner countries (together known as OPEC+) as a part of Declaration of Cooperation [22] has significantly impacted the recovery rate in various disasters in the recent years.

On a positive note, oil demand will stay in existence for at least the next 20 years until 2040 (although at a slower pace), as reported by British Petroleum's (BP's) World Energy 2019 Outlook [23], while demand for natural gas seems to be the limelight of energy resource growth in addition to renewable energy sources.

There has been an upsurge in the optimism for industry's performance in recent years though, due to several factors including consistent 3 years of recovery for oil prices in 2019 [24] and swift recovery in 2021 amidst the COVID pandemic [25], significant growth in upstream (discussed in the upcoming sections) as well as midstream sector, stabilization in crude prices ($50/barrel in 2019 after consistent recovery and $60/barrel in 2021 after COVID pandemic recovery), increase in job creation (boosting employment) of hundreds of thousands of personnel and steep increase in active drilling rigs (591 to 780+ in United States for the year 2019 alone). These factors put out a completely different and hopeful future for the industry.

Moreover, for many years, the industry faced the fear of "running out of oil"; however, in recent years, the fear has been more about "fear of running out of demand", due to technical advances, political regulations around oil consumption and environment emissions and pacing transition into the field of green revolution. The high oil prices have also paved the way for various industries to reduce their reliance on oil and gas, to offset their economic, political and social matters in the long term. Furthermore, the threat of introduction of cheaper alternatives, feasible oil substitutes and supply from shale low permeability formations, deep-water and pre-salt oil deposits is imminent as well. As an inevitable outcome, the industry has faced some turmoil in the recent decade, including fall in oil prices since 2014 of up to 70% [26], decrease in oil production by 3%–5% [7] and increase in mass layoffs and unemployment up to 40% [27]. However, the industry leaders still possess optimistic perspective about the future due to various factors including demand and supply consistency around the world, positive outlook on COVID forced changes and transition towards net-zero emissions [28–30] while providing increased dividends, excess cash flow and clean energy investments.

1.1.4   Industry knowledge

On a scientific level, oil and gas consist of hydrocarbons (molecules formed due to chemical reactions of two basic elements – hydrogen and carbon) as their core element, which are naturally occurring substances found in rocks in the earth's crust. They are created by the decomposition and compression of plant and animal remains in sedimentary rocks (such as sandstone, limestone and shale). The formation of hydrocarbons occurs as a result of sedimentary rocks' creation due to ocean and water bodies' deposits along with the integration of remains of dead plants and animals into the rock formation. Eventually, oil and gas are generated once the pressure and temperature reach a certain threshold, deep in the ocean bed. Moreover, as the density of both oil and gas is less than water, they leak out of the bedrocks (if they are porous) or form an oil and gas reservoir (which tend to be the main sources of oil and gas production for most of the production companies in the industry). The drilling companies suck out the hydrocarbons until its commercially viable, after which the crack of suction is plugged and abandoned, with declaration of the reservoir as a "dry hole").

The production of these hydrocarbons in the industry also has unique measurement system(s). For instance, the oil production is generally measured in barrels, where a barrel (bbl) equals 42 US gallons [31]. Thus, the production descriptions usually contain metrics such as bbl/day or bbl/quarter. Furthermore, few prefixes in quantity are also useful to know, including "M" for 1000 (a thousand) and "MM" for 1,000,000 (a million). Thus, it is usual to see MMbbl/day or MMbbl/quarter in various E&P companies' report(s).

An interesting aspect deducted after careful analysis of the industry's tremendous changes throughout the years is the fact that the industry goes through specific breakthroughs as each decade passes [32]. The 1980s decade introduced directional drilling (with parallel measurement), while the 1990s decade brought automated drilling into the picture. The following decade brought in 3D printing to the industry while the current decade has been about Virtual and/or Augmented Reality, with a bright prediction onto the future of drastic changes. The intricate detailed outlook of each decade along with its breakthroughs can be understood via Figure 1.3.

Petroleum also happens to be the world's most actively traded commodity. The international petroleum market comprises thousands of participants who help facilitate

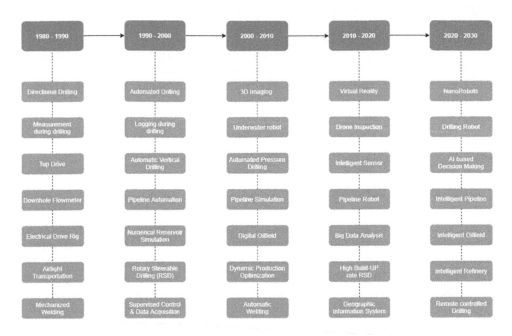

Figure 1.3 Decade-wise breakthroughs in petroleum industry.

the movement of petroleum from its origination point till its final destination to consumer (upstream to downstream). On a side note, "Brent" is considered as the international benchmark for petroleum pricing. However, that can change as per the region where the transactions are taking place as well. For instance, West Texas Intermediate or "WTI" benchmark is used in the United States while Dubai crude is used in the Gulf countries [33].

Moving onto the supply chain and processes involved in the life cycle of petroleum, the production is divided into three major parts, namely downstream, midstream and upstream [34]. The upstream segment consists of production aspects pertaining to exploration and production (E&P) of oil and gas. This includes the search for reservoirs and the drilling operation for oil and gas extraction. The companies associated in upstream segment are termed as E&P and are usually linked with high risks, high investment capital, extended process timeline and intensive technical outlook. Midstream segment, on the other hand, is concerned with the transportation of upstream products and thus involves the moving of raw extracted hydrocarbons to the oil and gas refineries. The companies associated with midstream segment take care of shipping, pipelines, trucking and raw materials' storage, while being linked with high regulations and low capital risk. Finally, at the end, the downstream segment is concerned with the refining process of the upstream raw materials brought through midstream transportation channel(s) for removing unwanted impurities through filtration (refining crude oil and purifying natural gas) and converting the raw materials to general public accessible products such as petrol, gasoline, lubricants, kerosene and liquefied petroleum gas among various others. The companies associated in the downstream segment include operations in all facets including oil refineries, natural gas distributors, petrochemical plants, petroleum product distributors and retail outlets. A closer look at these segments/phases of the petroleum production cycle will be undertaken in the upcoming sections.

As seen till now, the Oil and Gas industry has been an epitome of world economy, playing a major role in various global activities, while going difficult times throughout the years while maintaining its relevance till date despite aforementioned setbacks. The future of the industry still looks promising (as envisioned by industry leaders) due to its historical record of endurance and persistence, thanks to technological advancements, governmental bodies' pressure on transition and increased investment in clean & green emission technology for net-zero emissions.

1.2   PETROLEUM UPSTREAM

The first step in the complete petroleum lifecycle is the process of extraction of raw materials, hydrocarbons in our case, for further processing. This is where petroleum upstream comes in. It is also known as E&P segment, as it has major involvement in acquiring the raw hydrocarbons through various activities, including potential reservoir search, exploration, well drilling, oil extraction (or recovery), well operations and maintenance. It is to be noted though that the operations that occur after "search and exploration" activities only take place if the outcome of the extraction is deemed viable and profitable for the E&P companies.

There are two major types of resources (petroleum reservoirs), namely, conventional and unconventional resources. Conventional resources refer to the reservoirs that are traditionally found in deposits and traps (for which the required investments are standard). However, the unconventional resources generally refer to the source of the petroleum's generation (known as "the kitchen" in the industry), and the extracted hydrocarbons are termed as shale hydrocarbons. The investment required (cost and resource input for execution) in unconventional resources is usually extensive in nature; however, it has the capability to boost petroleum resources' output exponentially. Despite its potential, because it possesses high risk, early failure, continuous drilling requirement, cost inefficiency, multi-stage hydraulic fracturing and higher environmental damage and pollution risks, the utilization of unconventional resources has not reached its full potential as of now.

Political affairs do take place in each petroleum industry, and the upstream segment is no different. As most of the E&P companies lease out the natural resources from the states that own them, the host states do ensure that they receive a hefty amount of revenue from their lease agreement. Therefore, taking all these matters into consideration, there has always been tension in the balancing process of company profits, host state revenue, royalties and taxes owed to the government.

The world has generally seen an increase in production over each decade, and the production quantity has almost always exceeded prior estimates. For instance, the oil production from 1980 to 2011 of 795,000 MMbbl was considerably higher than the estimate of 683,000 MMbbl in 1980. There has also been a significant increase in production increase (of 30%) along with an increase of almost 100% in the amount of the remaining future production reserves. The same is the scenario with gas production, where the estimates for gas reserves increased more than 100% in the period of 1980–2011 timeframe compared to prior 1980 estimates, along with a similar (comparatively slightly less) increase in the quantity of actual gas production from those reserves [35].

1.2.1 Exploration

The upstream segment operations consist of three main parts, namely, exploration, drilling (or development) and production. The exploration part, as the name suggests, includes the exploration by search of hydrocarbons through several means (including geological research, geological and geophysical surveys and seismology). It is thus usually considered a high-risk and high-cost activity. Moreover, exploration stage also includes the leasing agreements and acquiring permission from owners of resources (thought to contain economically viable petroleum reservoirs). This process generally includes, but not limited to, the following steps: opening for tenders, international competitive bidding, bidding factors & award procedures, data purchase, defining of negotiable bid factors/terms, invitation to bid and direct negotiations (possibly followed by re-tender).

1.2.2 Drilling

The next step in the segment is drilling (or development) on the onshore and offshore sites that have been leased, proved to possess a desirable amount of resource and have proven to be economically viable. The drilling process (almost always) comes in some form of uncertainty, even after geological and geophysical surveys. Thus, the only method to attest certainty to the resource is to dig an exploratory well and actually test the viability of the resource at hand. The drilling of these wells (exploratory as well as final petroleum ones) is majorly taken care of by Oilfield Services sector companies. Furthermore, once the well is drilled, the appraisal and determination of commerciality is undertaken through various steps portrayed in Figure 1.4. These

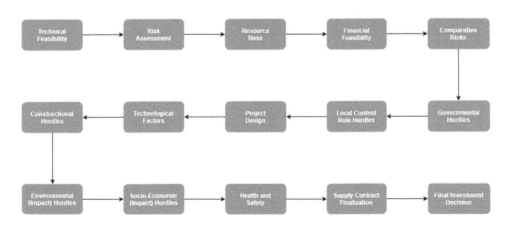

Figure 1.4 Appraisal and determination of commerciality in development.

qualitative and quantitative assessments help in measuring the feasibility of the project at a granular level with stark clear stats to make a constructive decision to move forward [36]. A successful development stage outcomes look something like this: formalized and achievable project phases, checkpoints in place, clear project implementation team's accountabilities and continuous cost control & risk monitoring.

1.2.3   Production

Once the exploratory well tests turn out to be successful, the final part (or stage) of the upstream segment is carried out, which is production. Production mainly deals with the actual extraction of resources along with bringing them to the surface and aims to maximize the amount of petroleum extracted from the chosen reservoir. The major activity in the production stage is the efficient recovery of hydrocarbons via primary, secondary and tertiary recovery methods. The primary recovery methods comprise of petroleum extraction through processes such as natural hydrocarbon rise to surface, pump jacks and/or other artificial lifting devices. They account for 5%–10% recovery of total reservoir capacity. The secondary recovery methods include air, water and/or other fluid injection into the reservoir to force the hydrocarbons to move up from their resting place by displacing them (accounting for additional 30% recovery). And finally, the tertiary recovery methods use various techniques (like thermal, gas and chemical injection into reservoirs) accounting for 15%–50% of total reservoir recovery. This is followed by plug and abandonment, which marks the end of the productive life of a well.

All the steps in the segment can be overviewed and grasped with the help of Figure 1.5, which portrays different aspects (timeline, sub-steps, etc.) of exploration, drilling or development and production.

Moreover, as described, the entire timeline of upstream processes can be loosely defined through the hereby mentioned stats. The initial exploration stage itself can take up to 5–7 years, while development, leasing and drilling accounting for two more years (which can overlap with the exploration stage) and dependent on several factors such as weather conditions, drilling depths, drilling rock hardness and site distance [37]. Once everything's considered and agreed upon with infrastructure in place, the production period can last around 20–30 years (with possible extensions up to 10 years if the commercial viability of productions remains viable), after which the production site is plugged and abandoned as mentioned earlier.

1.2.4   Segment finances

A universal aspect of any industry happens to be the proper management of expenditure for effective working of its activities while maximizing the profit margins. Therefore, cost management is a vital aspect to understand everywhere. In lieu of this, the major factors influencing the cost management in the upstream segment can be understood via Figure 1.6. The E&P companies tend to majorly balance out these factors in order to maximize output and profits.

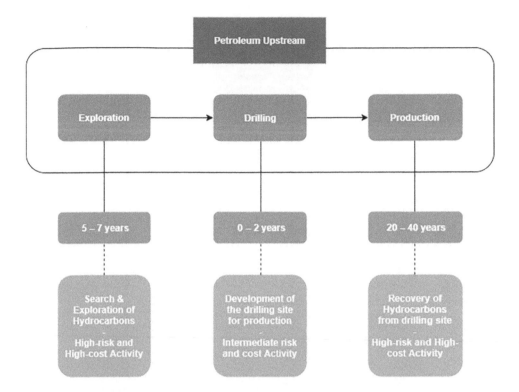

Figure 1.5 Petroleum upstream overview.

1.2.5 Segment knowledge

Few terms and terminologies that are widely used in the upstream segment include "independents" (companies that exclusively operate in upstream and are pure E&P companies), "IOCs" (abbreviation for Integrated Oil Companies, referring to companies that have assets in the midstream and downstream segment as well), "NOCs" (abbreviation for National Oil Companies, which refer to companies that are owned and managed by governments around the world), "NGL" (abbreviation for Natural Gas Liquids that consist of liquid hydrocarbons with natural gas such as methane, ethane, propane & butane), LNG (abbreviation for Liquefied Natural Gas that consist of gaseous portion of natural gas in a liquified state such as methane and a little ethane) and "Oilfield Services" (companies that build infrastructure and provide specialized assistance (equipment, services and skills) required for exploring, drilling, testing, producing, maintaining and reclaiming petroleum from the reservoirs).

Some of the major characteristics of petroleum reserve(s) are the presence of four key geological formations, namely, source rock, reservoir, seal and trap. The source rock refers to the location or source of petroleum generation (oil and gas generation). The reservoir is the place the generated petroleum resides (usually by travelling through microscopic porous structures in surrounding rock). A seal refers to a layer

Figure 1.6 Cost management in upstream segment.

around (usually above) the reservoir that is impermeable and stops further movement of petroleum. Finally, a trap refers to a combination of seals and reservoirs.

The industry also frequently uses the terms like "'proven" reserves. These proven reserves refer to the measure of a company's extent of belief in the production of economically recoverable hydrocarbons from a test site using existing technology at that point of time along with existing economic and operating conditions. The measure is generally updated constantly throughout the entirety of the upstream segment's activities (based on consistent re-assessments on scheduled intervals of time). There can be gigantic differences in the measurements taken in re-assessments, mostly due to advances in technology. For instance, Marcellus Shale's proven reserve measurements were increased by a factor of 40 by U.S. Geological Survey due to the advances in hydraulic fracturing and horizontal drilling [38]. Furthermore, beyond proven reserves, "probable" resources also exist that have 50% or higher chances of meeting the feasibility criteria for a potential reservoir, and finally, "possible" reserves exist as well that haven't yet been confirmed to be feasible (but there's a possibility). The recovery rate (production from reservoir) is generally 90% in case of proven reserves while it drops to 50% in case of probable ones (with no certainty in possible reserves). However, these standard values are subject to significant change due to various factors including reservoir quality and consistency, field production strategies, well & reservoir fluid properties and other geological factors (like rock permeability, porosity and water saturation).

Moving forward with the segment's classification aspects, the raw petroleum produced from drilling sites, often called crude oil, comes in two major types – light (or sweet) and heavy (or sour). Light crude oil has low density, low viscosity, low specific gravity, high API gravity (measurement of how heavy/light a petroleum liquid is compared to water), flows freely at room temperature and generally has low wax content. The heavy crude oil, on the other hand, has high density, high viscosity, low API gravity (less than 20°, with extra heavy oil possessing less than 10° value), does not flow easily and has comparatively higher wax content. Moreover, as light or sweet crude oil provides higher percentage of gasoline and diesel in the refining processes of the downstream segment, it is usually sold at a premium price in global trade. The presence of sulphur and other impurities in heavy or sour crude oil, along with its comparatively higher negative effects on the environment, also makes it a less viable product as it drives up the refining costs in the downstream segment (further making light crude/sweet oil better economically). [Few standard benchmarks utilized globally for classifying crude oil include Light Louisiana Sweet, West Texas Intermediate and Petrozuata Heavy.]

Due to its inherent nature of activities, the upstream segment is majorly categorized as high risk–high return segment due to enormous risks and heavy investment ($5–$20 million per site for exploration alone) is associated with it along with substantial profits in case everything goes well. It is also highly regulated by nations' policies due to several factors mentioned before, which significantly impact the entire supply chain. Moreover, the segment is also heavily impacted by global politics (or state affairs) including political instabilities (like war, and/or international conflicts), laws & regulations (like environmental restrictions & social programmes), state-imposed price controls, tax regimes and respect for contracts. Another major aspect of upstream happens to be the uncertainty of production and its outcome, as several factors (like supply & demand, economic growth, recessions, crude production quotas, seasonal weather patterns and severe weather events' disruptions) play a major role in the success of upstream operations. And the final characteristic of upstream resides in the intensive requirements of technology (technologically intensive) due to the complex processes involved in all aforementioned activities.

1.3 PETROLEUM MIDSTREAM

Petroleum midstream segment refers to the segment that operates between upstream and downstream. It comprises three major steps, namely, storage, processing and transportation of raw petroleum products extracted from the upstream segment. This segment takes input from the upstream segment (crude oil, natural gas and NGL) to provide services to the downstream segment (refineries associated with downstream distributors). It is therefore responsible for linking far-reaching petroleum-producing areas (in the upstream segment) and population centres (in the downstream segment) where refineries and final consumers are located.

The Transmission pipeline companies, storage warehouses and processing facilities are included in this segment as well. Therefore, the companies that operate in the midstream segment usually specialize in the operation of tanker ships, pipelines (by pipeline, rail, oil tanker(s) or truck) and/or storage facilities. Each step can be seen in Figure 1.7 that has summarized the important aspects for the same.

1.3.1   Processing

The processing step, also known as the refining step, turns crude oil (raw extracted carbons comprising a mixture of oil, natural gas, and NGL) into marketable products like heating oil, gasoline, jet fuel and diesel oil among other commercially viable fuels for end consumers. This can take place on-site (mostly) or off-site (irrespective of whether the reservoir is onshore or offshore). Once the raw crude oil is obtained, it's passed from an oil-gas separation unit that separates water & oil from gas inside the complete mixture.

Thereafter, all the separated components are shipped to different locations through separated mediums (usually pipelines) as discussed in the transportation step. On a side note, the water can be recycled, and oil and gas refined before they are shipped as well. Some of the major oil refining processes include distillation, vacuum distillation,

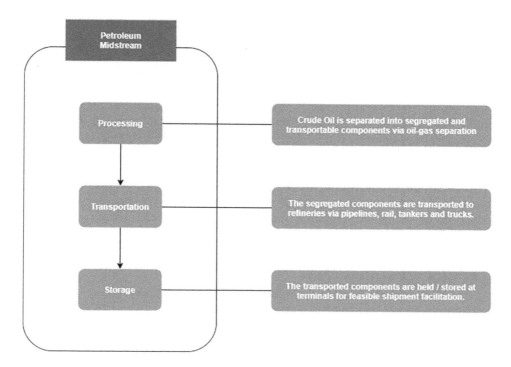

Figure 1.7 Petroleum midstream overview.

catalytic reforming and/or cracking, alkylation and isomerization, while gas refining processes include compression, glycol dehydration, amine treating and fractionation.

This step therefore helps in refining (removing waste), measuring (of product during transit and storage on a consistent basis) and compressing (breakdown into simpler and condensed transportable products) for feasible transportation to end users (power plants, export facilities, factories, gas stations, petrochemical product manufacturing plants and households) in the downstream segment. Furthermore, it also includes temporary storage of the hydrocarbons (for later transportation), measurement of the quantities (production rate or total production of hydrocarbons) and any wholesale marketing efforts to consumers that are required for the operations' success.

1.3.2   Transportation

The transportation step takes care of the transfer of crude oil to refining facilities (and then to end users as well) through different means (like pipelines, rail, tankers and trucks). Pipelines are generally the most suited means of transportation due to their cost effectiveness, long-distance feasibility (even across continents) and substantial safety. The pipelines used for oil transportation are usually small-diameter (comparatively) field gathering pipelines, while the ones used for gas transportation are large–diameter high pressure–handling pipelines called transmission lines (as gas flows at much higher pressure than crude oil). Although pipelines are capital intensive in the beginning, they have relatively low operational costs (for a considerable long period of time, i.e., lifetime of pipelines) once infrastructure is in place.

Tanker ships and barges are also useful in case of long-distance transportation (including international transportation). However, their usage is generally not preferred over pipelines due to environmental impacts in case of mishaps like oil tanker spills that destroy flora and fauna in the surrounding oceanic bodies. For instance, there were 358 spills in the 1990s totalling 1,133,000 tonnes of lost oil, 181 spills in the 2000s totalling 196,000 tonnes and 42 spills in the first half of the 2010s totalling 33,000 tonnes of lost oil [33]. This puts a major concern to environmentalists, legislators and investors (making pipelines a much-preferred option).

Rail and truck are also used for moving petroleum products (comparatively lowest volume) but are generally preferred for shorter routes because of their cost-effectiveness and inherent compensation for inflexibility of other major transportation means like pipelines and ships. In essence, pipelines and tankers are preferred for long-distance destination, rail or barge for intermediate length destinations and trucks are usually preferred for short-distance destination transportation.

Such a variety of transportation means that it's always a trade-off between all the options to look for the one that's feasible at that point of time and given certain conditions. For instance, although pipelines are the safest and most efficient means of transportation medium, a new pipeline infrastructure can take a considerable amount of time and investment which can be a deal breaker in several scenarios. The political resistance can also be excruciating sometimes. On the other hand, although railways and trucks are not the most efficient or safest means of transport, they are flexible in both timing and destination (and thus prove to be useful in scenarios where time and/ or destination is a constraint).

1.3.3 Storage

The storage step is also vital to the entire midstream segment as it deals with providing storage facilities at terminals (also termed as tank farms, oil installations or oil depots that are locations connecting various pipelines where raw materials can be stored) throughout the petroleum distribution systems. These terminals are usually connected to pipeline systems for feasible shipment facilitation and are often located near refining, processing plants or in-transit locations. Moreover, while oil-based products are held on surface storage tanks, gas is generally stored in underground facilities (such as commonly used depleted reservoirs, aquifers and/or salt dome caverns) for self-explanatory reasons like extremely high pressures.

Furthermore, other than companies, nations also actively participate in this step for mainly security purposes to avoid energy security for the proper functioning of their country. For instance, the strategic petroleum reserve currently stores approximately 727 million barrels (with wound 695 million barrels or 36 days consumption) in inventory and is United States Department of Energy's maintained emergency oil storage. It was created in 1975 after the 1973–1974 supply disruption's oil embargo [39] and serves as a response to any and all forms of disruptions, mishaps or threats to the US' energy crisis. Also, as the energy sector is closely related to the financial sector in any country, such storages by nations also ensure stability in their economy.

1.3.4 Segment knowledge

The major characteristics of the midstream segment comprise high regulations (due to catastrophic environmental consequences in case of mishaps) and low capital risk (due to its inherent non-risk profile of nature of operations and relatively less complexity of assets compared to other segments). Some of the major bodies concerning transportation include the European Union in Europe and Federal Energy Regulatory Commission in the US.

There are several negative environmental impacts that can be directly attributed to midstream segment activities too, which support the heavy regulations that are imparted on it. The most obvious impacts turn out to be the ones associated with transportation activities, which heavily rely on fossil fuels as key working ingredient. This results in severe effects on surrounding air quality along with the increase in greenhouse gas emissions. Another major issue that's inherently associated with downstream is the possibility of accidental leaks, spills and explosions caused by transportation medium (like pipelines, railroads and ship tankers). And lastly, the development of infrastructure necessary for proper function of the segment also imparts damage to natural wildlife habitats. In essence, midstream activities affect air pollution, water pollution, climate change and habitat loss.

Moreover, due to its unique positioning, it's usual to find various upstream and downstream elements in the midstream segment as well. For instance, it can include purifying plants (which raw natural gas passes through before entering the refinery). This also happens to be one of the reasons why several midstream companies are also IOCs.

The companies involved in midstream segment are mostly either integrated or independent oil companies. IOCs own assets in all the segments to maximize their profit, while independent oil companies provide services, equipment and expertise to other companies that require midstream assets. Major midstream companies include Koch, Transmontainge, Aux Sable, Bridger Group, DCP Midstream Partners, Enbridge Energy Partners, Enterprise Products Partners, Genesis Energy and Gibson Energy.

On a side note, it's necessary to note that the segment's success is dependent on some key factors including the success of upstream firms on continuous delivery of reserves, refinery margins that encourage refined product production, health of every segment's consumer markets (due to its positioning), petroleum price levels (that impact the profitability of operations) and finally, the political sentiment of states for pipeline expansion(s) (due to "not in my backyard"-based perspective hurdles)!

1.4 PETROLEUM DOWNSTREAM

The downstream segment of petroleum refers to the last and final segment pertaining to petroleum industry's operations. It comprises several activities including refining and processing of petroleum, supply and trading of petroleum product as well as marketing of products created as a result of the downstream activities to end consumers (that can be private businesses, national institutions, governments or private individuals). Several activities (if not all) in petroleum downstream segment overlap with midstream segment processes, and the separation between "what resides where", is usually blurred (to an extent where midstream operations are generally considered to be a part of downstream operation in various guides, literature, documentations and real-world practical reports).

Downstream segment is responsible for the creation of products that include (but not limited to) familiar items like petrol, diesel oil, kerosene, natural gas, liquefied petroleum gas, plastics & pharmaceuticals and unfamiliar items like jet fuel, heating oil, fertilizers, pesticides, lubricants, synthetic rubber, waxes & asphalt among thousands of different petrochemical-based products. Moreover, each activity can be seen in Figure 1.8 with summarized aspects for the same.

1.4.1 Refining and processing

A major part of entire downstream operations is petrochemical refining of raw crude oil, which is basically a mixture of thousands of different hydrocarbons (where each component has its own weight, density, size, texture and importantly, boiling temperature). These components are separated by the refining plants, and their finished product are categorized into three main types – light products (liquid petroleum gas, gasoline/petrol and naphtha), medium products (kerosene, jet/aircraft fuels and diesel fuel) and heavy products (lubricating oils, paraffin wax, petroleum coke and asphalt/tar). An elaborated comprehension of the refining of different components can be understood with the help of Figure 1.9.

Major international companies specializing in refining include Irving Oil, Petroplus Holdings, ERG Petroli and ENI, while major IOCs that undergo refining operations include British Petroleum, ExxonMobil, Chevron, Shell and Total.

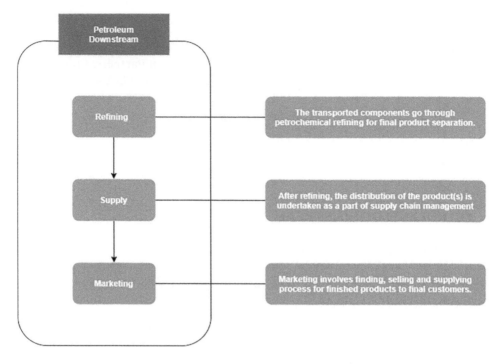

Figure 1.8 Petroleum downstream overview.

1.4.2   Supply and trade

The segment is also involved in the supply (distribution) and trade of the refined products that it produces through the first activity. The distribution is a major part of "supply chain management", which refers to the coordination of all the operations involved in the making & moving of (here, refined) products. The segment itself relies on the supply chain (consisting of businesses and their operations that utilize the products and supply them to end consumer) for distribution.

This step also takes care of finances for maximizing profits by making the actual trade of the products to the consumers through retail (or now even online e-commerce) outlets. Moreover, the changes in technology have made this part of operation less time consuming, more effective, risk-free and streamlined, which now reduces the total human involvement (that allows for us to focus and invest more time and resources into other activities).

1.4.3   Marketing and retail

Downstream segment also engages, as mentioned earlier, in the actual marketing of these products to the end consumers along with distribution related to the same to stores and companies specialized in selling. Product marketing is the business of finding, selling and supplying process for finished products to direct (like airlines, utilities,

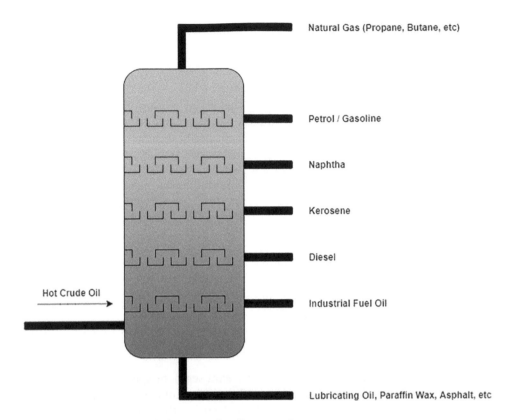

Natural Gas (Propane, Butane, etc)

Petrol / Gasoline

Naphtha

Kerosene

Diesel

Hot Crude Oil

Industrial Fuel Oil

Lubricating Oil, Paraffin Wax, Asphalt, etc

Figure 1.9 Refining through fraction distillation column.

municipalities, trucking fleets, petrochemical and industrial manufacturers) or indirect (like motor oil producers, home fuel oil suppliers, propane tank distributors and service stations suppliers) consumers.

Furthermore, downstream segment also consists of operations such as retail and distribution for the products that it markets; and in fact, has been a major aspect of downstream segment and the contribution of midstream segment has been decreasing over the years (if not decades).

1.4.4   Segment finances

The pure business aspect of downstream aspect resides in the margins and thus is also called a margin business. A margin business is characterized by its working where the profits are made by the amount of margin present between the price of finished marketable products sold to consumers (here, petrochemical products) and the price of raw materials (here, crude oil brought in from upstream or midstream segment).

The profitability of the downstream segment is also an interesting aspect to look at. While most IOCs and other segment companies get hurt by lower oil prices (say, due to oversupply of petroleum), the downstream segment and its companies benefit from it substantially. This happens due to the fact that when petroleum prices fall sharply, petrochemical products lag the petroleum prices, and as a result, the refining margins grow (making substantial profits from the same). The converse is true too, as an increase in petroleum prices may result in declined profit margins.

Furthermore, the level of margins or profits for downstream companies also vary and can be affected by seasons (petroleum production has seasonality as well), parallel crude prices (of course), supply & demand (at that point of time) and production cuts (by institutions like OPEC). All these factors, when aligned properly, have a viable potential to inherently change the amount of profitability by a factor of 5 to 6!

The global demand for refined products is enormous. For instance, oil-based products like petrol, diesel and jet fuel account for 40% of the total global demand (while covering almost 65% market in the US). Oil is highly consumed in North, South & Central America and the Middle East. Moreover, it is to be noted that while coal dominates in Asia Pacific regions, natural gas is still a leading fuel in Europe and Eurasia.

1.4.5   Segment knowledge

In essence, the downstream segment is anything that relates with the post-production of petroleum activities and is responsible for creating the finished product (along with the actual sale of it). It's usual to hear in the industry that the closer a company is to end-consumers (for providing petrochemical products), the further down the stream it is said to be!

The companies associated with the downstream segment specialize in oil refineries, petrochemical product distribution, petrochemical plants, natural gas distribution and finally, retail outlets. Moreover, as mentioned earlier, the majority of downstream companies diversify into all levels of petroleum production (due to no clear distinction in midstream and downstream segments as well as to maximize their profits). Major downstream companies include Marathon Petroleum, Saudi Aramco, British Petroleum, Total, Qatar Petroleum, Halliburton and Schlumberger among various others.

Moreover, the largest producers of petroleum products in the world are the US, the European Union, China, Russia and India in descending order. Furthermore, the international capacity for crude refining is forecasted to grow exponentially as well with the lead taken by China, followed by Southeast Asia, Latin America and the Middle East [40].

An interesting aspect of the downstream segment is the major by-product that it creates. It is sulphur. As the majority (if not the entirety) of petroleum is made up of hydrocarbons, it usually has sulphur-based compounds inside it. These sulphur-based compounds are a part of impurities that the downstream plants remove during refining processes. In this case, the process is hydrodesulfurization, which converts the said sulphur-based compounds into gaseous hydrogen sulphide (H_2S). Once the sulphur content is removed from petroleum in form of (majorly) hydrogen sulphide, it is then again converted into elemental sulphur. This sulphur quenches the needs for global sulphur demand; for instance, 7.3 million metric tonnes of sulphur was produced by the United States in the year 2020 from the petroleum industry (downstream segment) alone [41]. On a side note, other major by-products of the segment also include different plastics and foam.

The downstream segment does impact different industries including energy, chemical and retail industry which are obvious to comprehend. However, there are various industries that it indirectly affects. For instance, it also affects medical industry due to the fact that it helps in the production of pharmaceuticals and products & equipment required & used by medical professionals. Likewise, the downstream segment also plays a key role in the agricultural industry as well, due to production of pesticides and fertilizers (including fuel required for farming equipment's functioning). It also affects the beauty industry, as most of the cosmetic products utilized inside it are created through high-quality oils that are extracted through petroleum.

Alike other streams, downstream also has a significant impact on the environment due to its involvement in various complex and dangerous activities, along with the effects caused by the use of its finished products. For instance, the distribution activities heavily rely on fossil fuels that eventually release greenhouse gas emissions (like carbon dioxide, sulphur oxides and nitrogen oxides) into the atmosphere.

Moreover, as the distribution lies inside the downstream segment as well, the possibility of leaks is present too and present as a hazardous and tragic threat to the environment with detrimental effects on flora and fauna. Although downstream operations represent a fraction of hazards and emissions (less than 10%) of total environment threats compared to upstream and midstream operations (and thus comparatively lesser levels of regulations), it is surely not something to be overlooked. Furthermore, the retailing part also poses threats such as infamous leaks of volatile organic compounds, adding to downstream segment's overall environmental impact.

There has been a reduction in the interest of downstream segment by IOCs and major international players due to various factors mentioned in Figure 1.10. Although consistent low returns compared to other segments has been the major factor for decreased interest, other factors like intense competition, strong dealer bargaining powers and threat of environment liabilities have also contributed to the gradual shift in every actor's (IOCs, and international private companies) interest as well as refrain in any subsequent investment in the segment (even divest downstream operations to focus elsewhere).

There is a notable contrast picture here though; NOCs remain to invest and operate in the downstream segment. They actually happen to be the major players in the segment, accounting for the majority of market share. This is because NOCs have different aim and motivation compared to IOCs & private companies' purely profit-driven goals. As NOCs care about aspects like domestic energy supply security, nation's employment opportunities and promotion of industrialization, they still (and will) continue operations in the downstream segment. This is in part the reason why any major progress (in technological, investment or scalability fronts) in the future will likely be from the NOCs.

In the lieu of such issues, all the remaining companies in the downstream business are putting significant efforts into making tremendous changes of revamping their operation and focusing on excellence, in order to offset the aforementioned challenges. Some of the major revamps include efficiency improvements in the refining and other industrial facilities (to make the most out of the raw product, reduce time and decrease environmental impact), customization of non-fuel products (to cater to the needs of specific individuals or groups, as there's no "one fits all" product), optimization of feedstock supply (to sustain a long-term sustainable and cost-effective crude oil

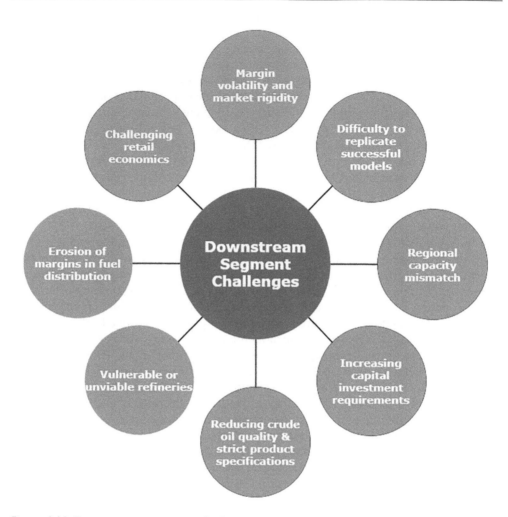

Figure 1.10 Downstream segment challenges.

supply) and integration of physical upstream segment (for safe supply and decreased exposure to crude price fluctuations), which are all summarized in Figure 1.11.

Furthermore, there are various financial-based strategical changes in business that can be undertaken, which may possibly provide successful transformation to downstream companies, and they are summarized in Figure 1.12.

1.5 BIG DATA

With the advent of technology and access to devices like smartphones, cheap information detection Internet of Things (IoT) devices, laptops, cameras, microphones, radio-frequency identification (RFID) readers, software logs, wireless networks, remote industrial, domestic and environmental sensors, the amount of data generated is humongous (on daily basis). The sheer increase in data generation can be understood

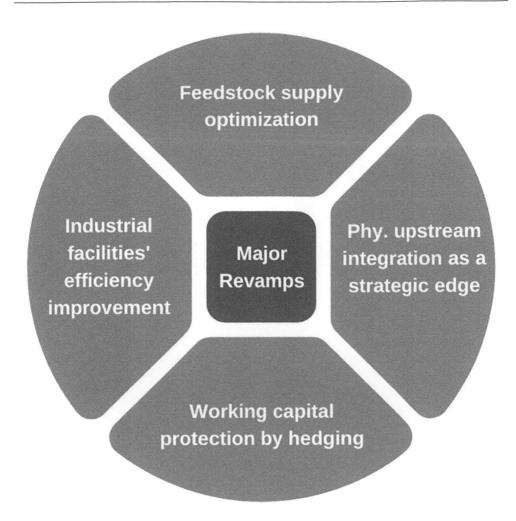

Figure 1.11 Downstream segment revamps.

by just realizing the fact that since 1980s, the capacity of the world to store (due to increase in its creation) has doubled every 40 months! [42] or that we will have produced 163 zetabytes (or 163 trillion gigabytes/GBs) of data by 2025 (as per the International Data Group report [43]).

Most of this data is huge, complex, unstructured and scattered collection of different data source datasets, and handling it is no easy feat (especially with traditional data application software or processing methods, who do not provide satisfactory results in acceptable amount of time). This is precisely where the concept of big data comes in.

Big data refers to a set of processes including data acquisition, data storage, search, share, transfer, analysis and visualization, which together aid in proper comprehension of the data at hand to make constructive decisions in real world. It can also be defined with the help of three major Vs, i.e., Volume, Velocity and Variety [44]. Volume refers to

Figure 1.12 Downstream segment financial revamps.

the amount or size of data (that can range from gigabytes to exabytes) along with other data structures correlated with it like transaction records, files and/or tables. Volume of big data has been on the rise of exponential explosion in the next decade.

Velocity refers to the methodology of transferring the data across devices, including near time, real time, batches and streams. It also comprises time and latency characteristics of data manipulation and handling. Variety, on the other hand, refers to the difference in various types of data formats in which big data exists, including structured, unstructured and semi-structured data [45] along with a combination of two or all (although it has been reported that almost 90% of all data generated is unstructured [46]). Major data formats in the world today are audio, images, video, documents, emails, text files and graphical data among several others.

Moreover, recent studies have brought more Vs into picture as well, like Value (uses cases or benefits of using big data for creating value), Variability (change in big

Figure 1.13 The 6Vs of big data.

data and its model(s) for feasibility) and Veracity (level/quality of the big data being used by the companies). The summarized information about all the six Vs can be comprehended with the help of Figure 1.13. Furthermore, there's one more aspect that a lot of recent studies have started taking into consideration in the last decade, which is complexity. Complexity refers to the level of difficulty for problem that the application of big data provides and is an important characteristic that should be given vital focus.

As perceived from the Vs, it can be noted that the main goal of big data is to equip entities using it to process and analyse (using various big data tools) the complex data with feasible amount of time and resources, which would otherwise be impossible to accomplish with traditional methods. Furthermore, as discussed previously, big data is a vast field and has numerous tools under its belt for the same that can be used to perform big data operations.

1.5.1 Big data tools

Big data is a vast field, and over the years, there has been a tsunami of tools and technologies that help in performing big data operations. However, some of them that stand out include Apache Hadoop, Apache Spark, Apache Storm, MongoDB, Apache Cassandra and R package (summarized in Figure 1.14).

Apache Hadoop happens to be one of the most widely used big data tool in every industry and is Apache's open-source framework that stores, processes and analyses data while running on commodity software. It was written in Java and supports parallel processing for enhanced performance. It also runs on cloud platforms and consists of four main parts – Hadoop Distributed File System (abbreviation for HDFS, a file distribution service with high scale bandwidth), MapReduce (big data processing's programming model), YARN (platform for scheduling and managing Hadoop resources) and Libraries (modules to make Hadoop work properly).

Apache Cassandra is another popular big data tool and is a distributed database that helps manage big data with high availability as well as scalability without compromising performance efficiency. It also happens to be one of the best tools to accommodate and work upon all (structure, unstructured and semi-structured) types of data. Furthermore, it also provides capabilities (characteristics) that no other database (even NoSQL) provides, namely, continuous availability of the data source, linear scalable performance, cloud availability points, simple operations, easy data distribution across data centres and scalable performance. It therefore provides ACID (Atomicity, Consistency, Isolation and Durability) properties to the entities using it.

MongoDB is an open-source data analytics tool that is widely used in big data, as it's a NoSQL database that provides cross-platform compatibility support, and thus is preferred by companies that require fast growth and have limited resources and time. It's written in JavaScript, C and C++ programming languages and can run on MEAN software stack, NET applications as well as Java platform. It is usually preferred for handling unstructured data (or data that changes frequently) and preferred in cloud infrastructures when flexibility is the key requirement (along with its cost-effective functioning).

Apache Spark, generally considered as the successor of Apache Hadoop, is a vital big data tool rapidly in increased use today. It takes care of the major drawbacks in Apache Hadoop like support for both real-time and batch processing. It also supports

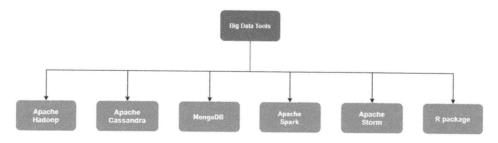

Figure 1.14 Big data tools.

in-memory calculation that inherently makes it 100 times faster (by reducing the amount of disk's read/write operations) than Apache Hadoop. It also provides versatility and flexibility, due to its capability to work with other tools such as HDFS and Apache Cassandra.

Apache Storm is another major open-source big data tool that's a distributed real-time and fault-tolerant processing system. It supports reliable processing of unbounded streams of data (data that's ever-growing, has a beginning but no defined end). It's written in Clojure (a dialect of Lisp programming language on the Java platform), and some of its major advantages of Storm include support for multiple language workability, support for JSON format protocols, sheer fast execution speed (million iterations per single second) and incredible scalability capability (while being fault-tolerant).

And finally, a major upcoming big data tool in the last decade has been the R package. It's one of the most comprehensively written statistical analysis packages out there. It was written in Fortran, C/C++ and R programming language, and is an open-source and multi-paradigm package that offers dynamic software environment. Although it is comparatively less efficient in security, memory management and speed, it is gaining popularity because of its vast package libraries (especially its unparalleled visualization aspects) while being free to use.

1.5.2   Big data work flow

As discussed earlier, big data comprises several activities including collection, ingestion, discovery, cleaning, integration, analysis and delivery that represent different aspects and work upon complete big data workflow for effective comprehension leading to productive decision-making. Figure 1.15 portrays the complete workflow for big data in a single glance.

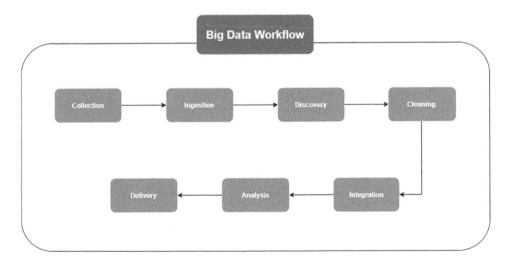

Figure 1.15 Big data workflow.

Collection refers to the acquisition of all types of data from various sources, while ingestion refers to the storing of the collected data into an appropriate data storage place (or "data store"). Discovery takes care of understanding the format and content of the data, while cleaning represents the cleaning process of data (formatting, structuring and segregating). The integration step comprises integrating the cleaned data (by entity extraction, entity resolution, indexing and/or data fusion). Analysis represents all the activities that pertain to understanding the data by performing operations on it (like Artificial Intelligence (AI), Statistics and Machine Learning among others), while delivery pictures the end of big data workflow with visualization, result showcase and deployment on actual company servers. All these parts processes make up an ideal workflow of an ideal implementation of big data in any industry.

1.5.3   Big data applications

Due to its inherent advantages, big data is utilized extensively in various industries. For instance, it is used in the banking industry for complex and massive transaction data processing as well as in the agriculture industry for real-time sensor data analysis (along with dynamic changing of environment using IoT devices based on data analysis). Moreover, it's also used in real estate for property analysis, better trend analysis and customer & market behaviour understanding. Healthcare industry has also seen massive usage of big data for effective treatment, providing personalized medicine, prescriptive analytics, waste reduction, automated external and internal reporting of patient data as well as rapid medical research. It's also used in Telecommunication for better customer experience, dynamic pricing, fraud detection and approximate churn prediction.

Insurance industry has also started using customers' big data like food, marital status, clothing size, media consumption and purchasing habits to make predictions on as to how much would their insurance cost (as per potential issues based on predictive analysis, although, it's been under some controversy [47]). Governments have also been relying heavily on big data to make their functioning efficient in terms of cost, productivity and innovation (for instance, NSA, in the US). Interestingly, education industry also has major applications of big data in two major ways. The first is its integration into the academic processes for gaining useful insights from the data analytics and providing personalized experience to students. The other one is the creation of specialization by reputed institutions (like University of Pittsburgh, Carnegie Mellon University, Johns Hopkins University, Missouri University of Science and Technology, University of California-Berkeley, University of Tennessee and UC Berkeley) to meet the demand for 1.5 million job vacancies for data professionals and managers as reported by McKinsey Global Institute [48]. Moreover, an elaborated list of big data applications comprising all applications mentioned until now is provided in Figure 1.16.

However, one less famous field where its use has been on the rise for decades and is exponentially been integrated into the respective industry happens to be Oil and Gas Industry.

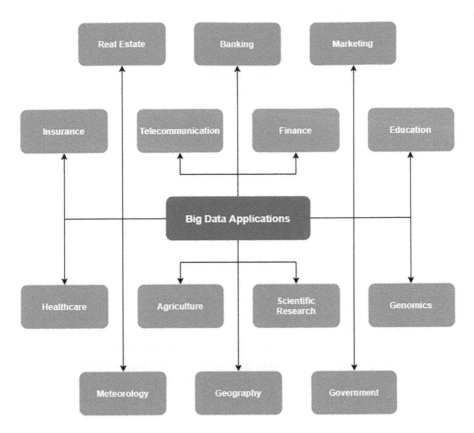

Figure 1.16 Big data applications.

1.5.4 Big data in petroleum streams

In continuity to the technological impacts by Oil and Gas industry mentioned in the earlier sections, some of the key technologies that pose huge benefits to all the segments include IoT, AI & Machine Learning, Wireless Communication, Augmented Reality, cybersecurity, blockchain, 3D printing and Robotics among various other. A key thing to note here is the dependence of each and every aforementioned technology on a single aspect of digitization – Data – for which Big Data enters the study of Oil and Gas industry due to its importance and core applicability to almost every facet of technological integration.

It is to be noted though that despite the potential that big data possesses, majority of O&G industries haven't quite grasped it properly to unleash its potential in the industry to help them. For instance, 70% of all the United States petroleum companies are not familiar with big data technology and the benefits it poses to their respective segment (or entire industry in case of IOCs) [49]. A report also stated that over half of the time of petroleum engineers and geoscientists is spent on just searching and assembling data [50]! Another study by Accenture and GE reported that over 80% of top executives from the petroleum industry considered big data to be one of the top three priorities in the upcoming years [51] to boost efficiency in the production and gas exploration aspects of the supply chain.

Big data, thus, has now started to make its way into the industry due to the massive amount of data generated by each operation in the supply chain (as shown in Table 1.3) [7], whose analysis can prove to be tremendously beneficial in the long term. The value of global petroleum industry for big data alone is expected to cross $11 billion mark by 2026 [52].

As perceived from the sheer amount of data generated in the petroleum industry from Table 1.3, it cannot be handled properly due to sustainability, analytical and storage issues [53]. This brings up the fact that although the humongous piles of data reside in O&G companies' archives and more data is generated each day, they literally just do not have the capability to analyse and process it [54]. This happens to be one of the core reasons for big data's rapid growth and integration into the industry.

Moreover, it is obvious to note that big data can aid several aspects of petroleum industry's operation including exploration (by analysing seismic data acquisition to develop 2D/3D images of the subsurface layers for potential sites' discovery), drilling (consistently improving the parameters pertaining to drilling such as speed, pressure and off-time, based on data generated by the drilling equipment in real time, making it efficient and performance driven [55]), production (by optimizing each activity associated with it), transportation (by solving issues like trade route and equipment optimization, based on data analytics), marketing (using AI and machine learning predictive analysis of consumers based on transactional data) and retailing (with behaviour analysis of consumers based on retail stores' big data). There are other operations as well that benefit from big data like reservoir engineering, refining and risk management, which will be thoroughly discussed in the upcoming chapters. Furthermore, big data can be classified into various categories in terms of petroleum industry's operations and is showcased in Figure 1.17 [49]. There are different intricate applications pertaining to big data when it comes to different operations associated with the supply chain present as well, which are portrayed in Table 1.4.

In essence, it is evident that big data has huge potential to transform the petroleum industry in almost all of its operations (as portrayed in previous paragraphs) and can be potentially applied to each segment for viable changes as shown in Table 1.4. A deeper look into each of them will be undertaken in upcoming chapters, covering several intricate aspects related to all of them!

Table 1.3 Petroleum industry data generation

Section	Data size
Drilling data	0.3 GB per well per day
Wireline data	5 GB per well per day
Fibre-optic data	0.1 GB per well per day
Seismic data	100 GB per survey
Plant process data	4–6 GB per day
Pipeline inspection	1.5 TB per 600 km
Plant atmospheric data	0.1 GB per day
Plant operational data	8 GB per year
Vibration data	7.5 GB per year per customer

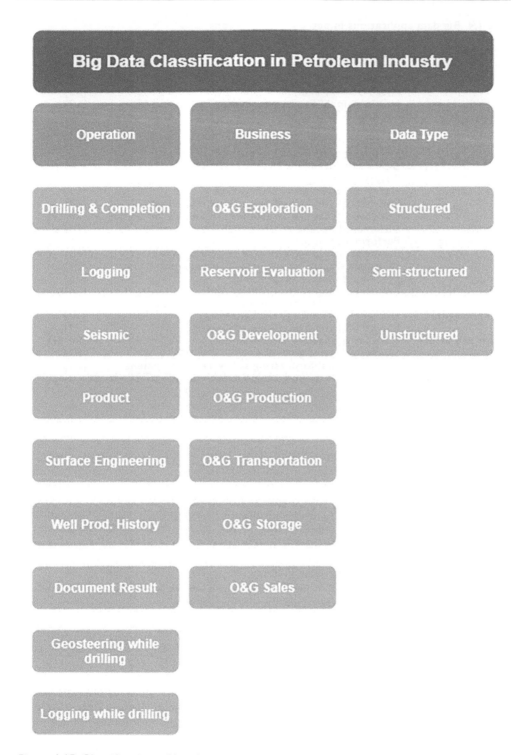

Figure 1.17 Classification of big data in the petroleum industry.

Table 1.4 Big data applications in petroleum operations

Operation	Application	References
Exploration	Seismic analysis	[56,57]
	Micro-seismic analysis	[58]
	n-Dimensional geological maps	[59]
Drilling	Drilling rig efficiency	[60]
	Drilling performance	[61]
	Invisible non-production time	[62]
	Drilling operations risk reduction	[63]
	Drill string dynamics characterization	[64]
Reservoir engineering	Reservoir management	[65]
	Heavy oil reservoirs optimization	[66]
	Hydraulically fractured reservoirs' modelling	[67]
	Unconventional O&G resources' reservoir modelling	[68,69]
Production engineering	Automated decline analysis	[70]
	Performance optimization of ESPs	[71,72]
	Performance optimization of rod pump wells	[73]
	Hydraulic fracturing projects' improvement	[74]
Refining	Petroleum asset management	[75]
	Refinery management optimization	[76]
	Well completion parameters' workflow study	[77]
Transportation	Shipping performance optimization	[78]
	Energy efficiency model development	[79]
Health and safety	Occupational safety improvement	[80,81]
	Safety predictive analytics	[82]
	Hazard events' forecasting software development	[83]

REFERENCES

1. IBISWorld (2021). Global Biggest Industries by Revenue in 2021. https://www.ibisworld.com/global/industry-trends/biggest-industries-by-revenue/.
2. Kalyani, D. (2018). Oil Drilling & Gas Extraction in the US, IBISWorld Industry Report, 2018.
3. Granitz, E., & Klein, B. (1996). Monopolization by "raising rivals' costs": The standard oil case. *The Journal of Law and Economics*, 39(1), 1–47. https://doi.org/10.1086/467342.
4. Mitchell, J., Marcel, V., & Mitchell, B. (2012). *What Next for the Oil and Gas Industry?* Chatham House, London, 1–128. https://www.researchgate.net/publication/292958223_What_Next_for_the_Oil_and_Gas_Industry.
5. World Economic Forum (2017). Digital Transformation Initiative Oil and Gas Industry. http://reports.weforum.org/digital-transformation/wp-content/blogs.dir/94/mp/files/pages/files/dti-oil-and-gas-industry-white-paper.pdf.
6. Handscomb, C., Sharabura, S., & Woxholth, J. (2016). The Oil and Gas Organization of the Future. https://www.mckinsey.com/industries/oil-and-gas/our-insights/the-oil-and-gas-organization-of-the-future.
7. Lu, H., Guo, L., Azimi, M., & Huang, K. (2019). Oil and gas 4.0 era: A systematic review and outlook. *Computers in Industry*, 111, 68–90. https://doi.org/10.1016/j.compind.2019.06.007.
8. Renewable Power Generation Costs In (2019). e International Renewable Energy Agency (IRENA). https://www.irena.org/-/media/Files/IRENA/Agency/Publication/2020/Jun/IRENA_Power_Generation_Costs_2019.pdf.
9. The White House: President Obama Launches Advanced Manufacturing Partnership (2011). https://www.nist.gov/news-events/news/2011/06/president-obama-launches-advanced-manufacturing-partnership.

10. Executive Office of the President of the United States, A Framework for Revitalizing American Manufacturing (2009). https://obamawhitehouse.archives.gov/sites/default/files/microsites/20091216-maunfacturing-framework.pdf.

11. Executive Office of the President National Science and Technology Council, A National Strategic Plan for Advanced Manufacturing (2012). https://www.energy.gov/sites/prod/files/2013/11/f4/nstc_feb2012.pdf.

12. Executive Office of the President of the United States, Strategy for American Leadership in Advanced Manufacturing (2018). https://trumpwhitehouse.archives.gov/wp-content/uploads/2018/10/Advanced-Manufacturing-Strategic-Plan-2018.pdf.

13. Kania, E. B. (2019). Made in China 2025, Explained. https://thediplomat.com/2019/02/made-in-china-2025-explained/.

14. IANS (2017). China, Germany Pledge Closer Innovative Partnership. https://www.financialexpress.com/world-news/china-germany-pledge-closer-innovative-partnership/698033/.

15. Foresight: The Future of Manufacturing: A New Era of Opportunity and Challenge for the UK (2013). https://assets.publishing.service.gov.uk/government/uploads/system/uploads/attachment_data/file/255923/13-810-future-manufacturing-summary-report.pdf.

16. Lasi, H., Fettke, P., Kemper, H. G., Feld, T., & Hoffmann, M. (2014). Industry 4.0. *Business & Information Systems Engineering*, 6(4), 239–242. https://doi.org/10.1007/s12599-014-0334-4.

17. Federal Ministry of Education and Research, High-Tech Strategy 2025 for Germany (2021). https://www.bmbf.de/en/high-tech-strategy-2025.html.

18. Monodzukuri (2014). Summary of the White Paper on Manufacturing Industries. https://www.meti.go.jp/english/policy/mono_info_service/overall/index.html.

19. Mitra, N. K., Bravo, C. E., & Kumar, A. (2008). Revival of Mumbai High North – A Case Study. All Days, 1–6. https://doi.org/10.2118/113699-ms.

20. Hanlon, C. (2013). The Usumacinta Disaster. https://journals.library.mun.ca/ojs/index.php/prototype/article/view/443.

21. Yin, F., Hayworth, J. S., & Clement, T. P. (2015). A tale of two recent spills – Comparison of 2014 Galveston Bay and 2010 deepwater horizon oil spill residues. *PLoS One*, 10(2), 1–17. https://doi.org/10.1371/journal.pone.0118098.

22. Declaration of Cooperation (2016). Organization of the Petroleum Exporting Countries (OPEC). https://www.opec.org/opec_web/en/publications/4580.htm.

23. British Petroleum (2019). BP Energy Outlook. https://www.bp.com/en/global/corporate/news-and-insights/press-releases/bp-energy-outlook-2019.html.

24. DiChristopher, T. (2018). Wall Street Sees a Recovery for Oil Prices in 2019, but Lots of Risk. https://www.cnbc.com/2018/12/21/wall-street-sees-to-recovery-for-oil-prices-in-2019-but-lots-of-risk.html.

25. Nagle, P. (2021). The Oil Market Outlook: A Speedy Recovery. World Bank Blogs. https://blogs.worldbank.org/opendata/oil-market-outlook-speedy-recovery.

26. Depersio, G. (2019). Why Did Oil Prices Drop so Much in 2014? https://www.investopedia.com/ask/answers/030315/why-did-oil-prices-drop-so-much-2014.asp.

27. Hiller, J. (2019). U.S. Shale Producers Turn to Jobs Cuts as Investor Pressures Mount. https://www.reuters.com/article/us-usa-shale-job-cuts/u-s-shale-producers-turn-to-jobs-cuts-as-investor-pressures-mount-idUSKCN1RL2Q4.

28. ExxonMobil Annual Report (2020). https://corporate.exxonmobil.com/Investors/Annual-Report.

29. Shell Annual Report (2020). https://reports.shell.com/annual-report/2020/.

30. Chevron Annual Report (2020). https://www.chevron.com/annual-report.

31. Gorelick, S. M. (2010). Oil Panic and the Global Crisis. https://www.researchgate.net/publication/230554591.

32. Yang, J. H., Qiu, M. X., Hao, H. N., Zhao, X., & Guo, X. X. (2016). Intelligence – Oil and gas industrial development trend. *Oil Forum*, 35, 36–42. http://www.sykjlt.com/EN/10.3969/j.issn.1002-302x.2016.06.008.

33. Petroleum Industry Overview – Midstream. Extractives Hub and CEPMLP (Centre for Energy, Petroleum and Mineral Law & Policy, University of Dundee). https://www.extractiveshub.org/topic/view/id/30/chapterId/325.

34. Lu, H., Huang, K., Azimi, M., & Guo, L. (2019). Blockchain technology in the oil and gas industry: A review of applications, opportunities, challenges, and risks. *IEEE Access*, 7, 41426–41444. https://doi.org/10.1109/access.2019.2907695.

35. British Petroleum (2012). BP Statistical Review of World Energy, June 2012. https://www.laohamutuk.org/DVD/docs/BPWER2012report.pdf.

36. Petroleum Industry Overview – Upstream (2021). Extractives Hub and CEPMLP (Centre for Energy, Petroleum and Mineral Law & Policy, University of Dundee). https://www.extractiveshub.org/topic/view/id/30/chapterId/324.

37. Bhattacharyya, S. (2009). Oil and Gas Exploration and Production: Reserves, Costs, Contracts. *International Journal of Energy Sector Management*, 3(2), 220–222. https://doi.org/10.1108/17506220910970614.

38. Gecan, R., Tawil, N., & Lasky, M. (2015). The Economic and Budgetary Effects of Producing Oil and Natural Gas from Shale. https://www.cbo.gov/publication/49815.

39. Kilian, L. (2005). The Effects of Exogenous Oil Supply Shocks on Output and Inflation: Evidence from the G7 Countries. CEPR Discussion Paper No. 5404. https://ssrn.com/abstract=892873.

40. Petroleum Industry Overview – Midstream(2021). Extractives Hub and CEPMLP (Centre for Energy, Petroleum and Mineral Law & Policy, University of Dundee). https://www.extractiveshub.org/topic/view/id/30/chapterId/326.

41. Mineral Industry Surveys (2020). United States Geological Survey. https://www.usgs.gov/centers/nmic/sulfur-statistics-and-information.

42. Hilbert, M., & Lopez, P. (2011). The world's technological capacity to store, communicate, and compute information. *Science*, 332(6025), 60–65. https://doi.org/10.1126/science.1200970.

43. Reinsel, D., Gantz, J., & Rydning, J. (2017). *Data Age 2025: The Evolution of Data to Life-Critical*. Seagate, International Data Corporation, Framingham, MA. https://www.import.io/wp-content/uploads/2017/04/Seagate-WP-DataAge2025-March-2017.pdf.

44. Jain, A. (2016). The 5 V's of Big Data, Watson Health Perspectives, IBM. https://www.ibm.com/blogs/watson-health/the-5-vs-of-big-data/.

45. Sumbal, M. S., Tsui, E., & See-to, E. W. (2017). Interrelationship between big data and knowledge management: An exploratory study in the oil and gas sector. *Journal of Knowledge Management*, 21(1), 180–196. https://doi.org/10.1108/jkm-07-2016-0262.

46. Ishwarappa, J., & Anuradha, J. (2015). A brief introduction on big data 5Vs characteristics and hadoop technology. *Procedia Computer Science*, 48, 319–324. https://doi.org/10.1016/j.procs.2015.04.188.

47. Marshall, A. (2018). Health Insurers are Vacuuming Up Details about You – And It Could Raise Your Rates, Propublica. https://www.propublica.org/article/health-insurers-are-vacuuming-up-details-about-you-and-it-could-raise-your-rates.

48. Manyika, J., Chui, M., Brown, B., Bughin, J., Dobbs, R., Roxburgh, C., & Byers A. H. (2011). *Big Data: The Next Frontier for Innovation, Competition, and Productivity*. https://www.mckinsey.com/business-functions/mckinsey-digital/our-insights/big-data-the-next-frontier-for-innovation.

49. Feblowitz, J. (2012). The Big Deal about Big Data in Upstream Oil and Gas. https://www.semanticscholar.org/paper/The-Big-Deal-About-Big-Data-in-Upstream-Oil-and-Gas/74ef30cde5fd005f17f4e74add86b03a97d8f339.

50. Brulé, M. R. (2015). The Data Reservoir: How Big Data Technologies Advance Data Management and Analytics in E&P. Paper presented at the *SPE Digital Energy Conference and Exhibition*, The Woodlands, TX, March 2015. https://onepetro.org/SPEDEC/proceedings-abstract/15DEC/3-15DEC/D031S021R001/182442.

51. Sukapradja, A., Clark, J., Hermawan, H., & Tjiptowiyono, S. (2017). Sisi Nubi Dashboard: Implementation of Business Intelligence in Reservoir Modelling & Synthesis: Managing Big Data and Streamline the Decision Making Process. Paper presented at the *SPE/IATMI Asia Pacific Oil & Gas Conference and Exhibition*, Jakarta, Indonesia, October 2017. https://onepetro.org/SPEAPOG/proceedings-abstract/17APOG/2-17APOG/D021S012R002/194087.

52. Transparency Market Research (2018). Expenditure in IT Sector to Develop Tools for Data Analytics Contributes Majorly in Big Data Oil and Gas Market. https://www.transparencymarketresearch.com/pressrelease/big-data-oil-and-gas-market.htm.

53. Trifu, M. R., & Ivan, M. L. (2014). Big data: Present and future. *Database Systems Journal*, 5(1), 32–41.

54. Mohammadpoor, M., & Torabi, F. (2020). Big Data analytics in oil and gas industry: An emerging trend. *Petroleum*, 6(4), 321–328. https://doi.org/10.1016/j.petlm.2018.11.001.

55. Raymond, M. S., & Leffler, W. L. (2017). Oil and Gas Production in Nontechnical Language. https://books.google.co.in/books?id=JEw3DwAAQBAJ&newbks=0&hl=en&source=newbks_fb&redir_esc=y.

56. Roden, R., & Ferguson, J. (2016). Seismic interpretation in the age of big data. *SEG Technical Program Expanded Abstracts*. https://doi.org/10.1190/segam2016-13612308.1.

57. Alfaleh, A., Wang, Y., Yan, B., Killough, J., Song, H., & Wei, C. (2015). Topological Data Analysis to Solve Big Data Problem in Reservoir Engineering: Application to Inverted 4D Seismic Data. Day 3 Wednesday, September 30, 2015. https://doi.org/10.2118/174985-ms.

58. Joshi, P., Thapliyal, R., Chittambakkam, A. A., Ghosh, R., Bhowmick, S., & Khan, S. N. (2018). Big Data Analytics for Micro-Seismic Monitoring. Day 3 Thursday, March 22, 2018. https://doi.org/10.4043/28381-ms.

59. Olneva, T., Kuzmin, D., Rasskazova, S., & Timirgalin, A. (2018). Big Data Approach for Geological Study of the Big Region West Siberia. Day 1 Monday, September 24, 2018. https://doi.org/10.2118/191726-ms.

60. Duffy, W., Rigg, J., & Maidla, E. (2017). Efficiency Improvement in the Bakken Realized Through Drilling Data Processing Automation and the Recognition and Standardization of Best Safe Practices. Day 1 Tuesday, March 14, 2017. https://doi.org/10.2118/184724-ms.

61. Maidla, E., Maidla, W., Rigg, J., Crumrine, M., & Wolf-Zoellner, P. (2018). Drilling Analysis Using Big Data has been Misused and Abused. Day 2 Wednesday, March 7, 2018. https://doi.org/10.2118/189583-ms.

62. Yin, Q., Yang, J., Zhou, B., Jiang, M., Chen, X., Fu, C., Yan, L., Li, L., Li, Y., & Liu, Z. (2018). Improve the Drilling Operations Efficiency by the Big Data Mining of Real-Time Logging. https://doi.org/10.2118/189330-ms.

63. Johnston, J., & Guichard, A. (2015). New Findings in Drilling and Wells using Big Data Analytics. All Days. https://doi.org/10.4043/26021-ms.

64. Hutchinson, M., Thornton, B., Theys, P., & Bolt, H. (2018). Optimizing Drilling by Simulation and Automation with Big Data. Day 3 Wednesday, September 26, 2018. https://doi.org/10.2118/191427-ms.

65. Bello, O., Yang, D., Lazarus, S., Wang, X. S., & Denney, T. (2017). Next Generation Downhole Big Data Platform for Dynamic Data-Driven Well and Reservoir Management. Day 3 Wednesday, May 10, 2017. https://doi.org/10.2118/186033-ms.

66. Popa, A. S., Grijalva, E., Cassidy, S., Medel, J., & Cover, A. (2015). Intelligent Use of Big Data for Heavy Oil Reservoir Management. Day 2 Tuesday, September 29, 2015. https://doi.org/10.2118/174912-ms.

67. Udegbe, E., Morgan, E., & Srinivasan, S. (2017). From Face Detection to Fractured Reservoir Characterization: Big Data Analytics for Restimulation Candidate Selection. Day 3 Wednesday, October 11, 2017. https://doi.org/10.2118/187328-ms.

68. Lin, A. (2014). Principles of Big Data Algorithms and Application for Unconventional Oil and Gas Resources. All Days. https://doi.org/10.2118/172982-ms.

69. Chelmis, C., Zhao, J., Sorathia, V., Agarwal, S., & Prasanna, V. (2012). Semiautomatic, Semantic Assistance to Manual Curation of Data in Smart Oil Fields. All Days. https://doi.org/10.2118/153271-ms.

70. Seemann, D., Williamson, M., & Hasan, S. (2013). Improving Reservoir Management through Big Data Technologies. https://doi.org/10.2118/167482-ms.

71. Gupta, S., Saputelli, L., & Nikolaou, M. (2016). Big Data Analytics Workflow to Safeguard ESP Operations in Real-Time. https://doi.org/10.2118/181224-ms.

72. Sarapulov, N. P., & Khabibullin, R. A. (2017). Application of Big Data Tools for Unstructured Data Analysis to Improve ESP Operation Efficiency (Russian). *SPE Russian Petroleum Technology Conference*. https://doi.org/10.2118/187738-ru.

73. Palmer, T., & Turland, M. (2016). Proactive Rod Pump Optimization: Leveraging Big Data to Accelerate and Improve Operations. Day 1 Tuesday, October 25, 2016. https://doi.org/10.2118/181216-ms.

74. Betz, J. (2015). Low oil prices increase value of big data in fracturing. *Journal of Petroleum Technology*, 67(4), 60–61. https://doi.org/10.2118/0415-0060-jpt.

75. von Plate, M. (2016). Big Data Analytics for Prognostic Foresight. All Days. https://doi.org/10.2118/181037-ms.

76. Brelsford, R. (2018). Repsol Launches Big Data, AI Project at Tarragona Refinery. https://www.ogj.com/articles/2018/06/repsol-launches-big-data-ai-project-at-tarragona-refinery.html.

77. Khvostichenko, D., & Makarychev-Mikhailov, S. (2018). Effect of Fracturing Chemicals on Well Productivity: Avoiding Pitfalls in Big Data Analysis. Day 1 Wednesday, February 7, 2018. https://doi.org/10.2118/189551-ms.

78. Anagnostopoulos, A. (2018). Big Data Techniques for Ship Performance Study. Paper presented at the *The 28th International Ocean and Polar Engineering Conference*, Sapporo, Japan. https://onepetro.org/ISOPEIOPEC/proceedings-abstract/ISOPE18/All-ISOPE18/ISOPE-I-18-190/20389.

79. Park, S. W., Roh, M. I., Oh, M. J., Kim, S. H., Lee, W. J., Kim, I. I., & Kim C. Y. (2018). Estimation Model of Energy Efficiency Operational Indicator Using Public Data Based on Big Data Technology. Paper presented at the *The 28th International Ocean and Polar Engineering Conference*, Sapporo, Japan. https://onepetro.org/ISOPEIOPEC/proceedings-abstract/ISOPE18/All-ISOPE18/ISOPE-I-18-422/20578.

80. Tarrahi, M., & Shadravan, A. (2016a). Advanced Big Data Analytics Improves HSE Management. Day 1 Wednesday, April 20, 2016. https://doi.org/10.2118/180032-ms.

81. Tarrahi, M., & Shadravan, A. (2016b). Intelligent HSE Big Data Analytics Platform Promotes Occupational Safety. Day 3 Wednesday, September 28, 2016. https://doi.org/10.2118/181730-ms.

82. Pettinger, C. B. (2014). Leading Indicators, Culture and Big Data: Using Your Data to Eliminate Death. Paper presented at the *ASSE Professional Development Conference and Exposition*, Orlando, FL. https://onepetro.org/ASSPPDCE/proceedings-abstract/ASSE14/All-ASSE14/ASSE-14-689/78061.

83. Cadei, L., Montini, M., Landi, F., Porcelli, F., Michetti, V., Origgi, M., Tonegutti, M., & Duranton, S. (2018). Big Data Advanced Analytics to Forecast Operational Upsets in Upstream Production System. Day 2 Tuesday, November 13, 2018. https://doi.org/10.2118/193190-ms.

Chapter 2

Petroleum Operations

2.1 GEOSCIENCE

Geoscience, also known as Earth Science, is the study of Earth and includes so much more than rocks and volcanoes, as it comprises the processes that form and shape Earth's surface, the natural resources we use, and how water and ecosystems are interconnected with each other. It incorporates several scientific disciplines related by their applications to the study of the earth. In essence, it encompasses study and work with minerals, soils, energy resources, fossils, oceans and freshwater, the atmosphere, weather, environmental chemistry and biology and natural hazards among several others [1,2].

Given the wide array of aspects (pertaining to Earth) that it covers, it is obvious to note that the study also has a great deal of correlation to petroleum activities (which are primarily associated with Earth due to hydrocarbons). Moreover, due to the fact that it also deals with the study of physical and chemical constitution of Earth, Geoscience in the modern age also uses tools and techniques from other science fields as well, such as chemistry, physics, biology, math and computer science.

Moreover, once the concepts of geoscience are comprehended properly, it is important to note various aspects of them as well, such as hydrology, petrology and climatic conditions; all summarized in Figure 2.1.

These aspects will play a major role in the applications and benefits of big data (elaborated in the upcoming chapters). Furthermore, the role played by professionals in the field (also known as geoscientists) is a central one in various aspects of operations (mentioned in a while), majorly including climatic processes, volcanic activity, petrology and hydrology.

On a side note, on top of expertise and work pertaining to geoscientists, the engineers involved in the field have to perform several tasks [3] that briefly include:

* Climate and global process modelling
* Environmental remediation and engineering
* Petroleum and mining exploration and extraction
* Energy policy
* Natural hazards assessment
* Land use planning
* Land sciences

DOI: 10.1201/9781003185710-2

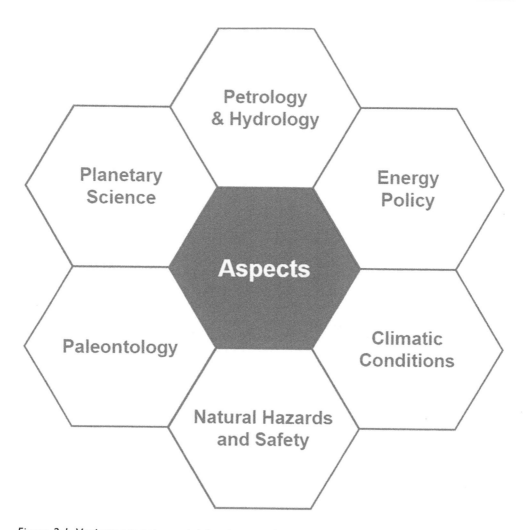

Figure 2.1 Various aspects pertaining to geoscience.

- Ocean sciences
- Planetary sciences
- Palaeontology
- Education

Now that the fundamentals have been cleared up, lets transition into the advent of big data into Geoscience. The wake of modernity and advancements in computer science & hardware have significantly skyrocketed the amount of data generated in geoscience. With such sheer amount of increase in data, traditional methods of data analysis fail to cope up [4], which is where big data comes in. Considering the fact that almost all operations associated with geoscience involve complex analysis for working of its key aspects, big data possesses massive potential for bringing benefits and applications like no other (discussed later in upcoming chapters). This is the result of insights that it provides through an input of cumulation of different data sources, by passing through big data analytics (using various tools and technologies).

Big data solutions help in collection, processing and analysis of data that geoscientists require to make their operations effective in every aspect (time, expenditure, efficiency, etc.). The data is collected through a network of sensors in various areas upon which big data analytics can be performed by geoscientists to develop strategic techniques of analysis for gaining meaningful insights (that could potentially save time and resources). These applications ultimately provide timely as well as actionable information about changes in the aforementioned areas, resulting in more control of companies over their operations along with increase in their operational outcome.

2.2 EXPLORATION AND DRILLING

Exploration and drilling are major operations in the upstream segment of petroleum production in oil and gas industry and mark a vital point in the entire workflow of industry's processes. Exploration refers to the activities (research, analyses, survey and seismology) done for the search of viable source of hydrocarbons and is deemed a high-risk operation that comes with massive expenditure. Drilling, on the other hand, is the operation that begins once a feasible source is located. As the name suggests, it refers to the actual drilling (or development) of sites (off-shore or on-shore) for the extraction of hydrocarbons underneath the surface. A general outline of appraisal and determination of decision for exploration and drilling operations can be seen in Figure 2.2.

Moreover, as the concepts of exploration and drilling are explained in detail in the first chapter, this section will majorly focus on big data's angle into the operations. Now, moving onto big data, as mentioned in the first chapter, it is on the verge of explosion in terms of usage and integration into almost each and every aspect of oil and gas industry's operations for benefits that range from efficiency improvements and profit maximization to reduced environmental impacts. And, the exploration and drilling operations are no different. On a stretch, its application is also termed as "the

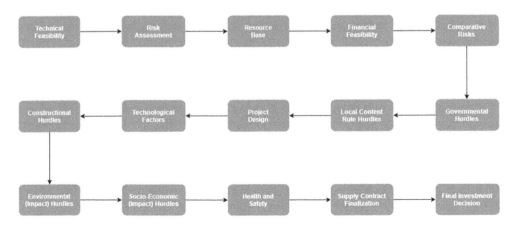

Figure 2.2 Appraisal and determination of commerciality of exploration and drilling operations.

dawn of the new era for the oil & gas industry" and "the fourth industrial revolution" by several experts.

The amount of data generated in exploration and drilling is simply gigantic and truly fits the notion of "big data". For instance, here are some statistics:

- A single well produces around 0.3 GB of data per day.
- A single exploration survey generates around 100 GB of data.
- The size of a single well's fibre-optic data happens to be 0.1 GB per day.
- The wireline data alone accounts for 5 GB per day for a single well.
- A single submersible monitoring pump generates 0.4 GB of data per day.

With such amount of data in the oil and gas industry, the big data revolution was bound to happen sooner or later. However, it is necessary to note that O&G companies hold a huge amount of what is termed as "dark data", which refers to data that organizations collect, process and/or store during standard business operations but tend to fail to use it for other purposes [5]. It basically refers to data that has value but is not exploited for it yet. And, this enormous pile of data is a potential gold mine for companies as the said big data analytics can prove to provide immense value to companies on top of its inherent value.

Also, while being considered as the oil of new economies [6], Big Data is utilized to access insights on only a mere 1% of its full potential (1% of entire data generated by each aspect of petroleum industry including exploration and drilling). This poses as a major opportunity to big data's implementation in oil and gas industry as even small improvements made by big data analytics can result in significant economic gains to O&G companies [7].

Moreover, even that big data was previously used majorly in seismic data analysis by using seismic interpretation software for a single reason of locating reservoirs that would otherwise be difficult to locate [8]. Although there has been an exponential increase in the investments made by O&G companies into information technology and analytics [9], their focus has generally been into innovation-orientated sectors such as tech sector and not into maximizing the potential of big data in their operations. However, with the exponential increase in the amount of data generated in exploration and drilling operations, traditional analysis tools and storage technologies now fail to cope up with managing, analysing and uncovering hidden patterns in the data [10,11]. This scenario has also resulted in a vital big data boom in the industry.

Following these aspects, it was noted in a study by Accenture and GE in 2018 that over 80% of top executives from the petroleum industry considered big data as one of the top three priorities of oil and gas companies in the upcoming years [12], which does not come as a surprise as petroleum industry is in a dire need of major revamps in terms of efficiency improvements and better decision-making process(es), especially in the exploration and drilling operations. Other than the business executives, this also helps the engineering team tremendously in terms of time reduction, as it has been stated that petroleum engineers and geoscientists spend around 50% of their entire work time to just collect and manage data (as per a report by Brule).

In a nutshell, the introduction of big data and data analytics along with digitization of Oilfield has literally brought ridiculous reduction in drilling time and operation costs in several regions by major O&G companies. For instance, during the low-price scenario in the North Sea (which is expected to account for the second largest share in the market), oilfield digitization reduced the operation costs by 40%, while ConocoPhillips saw an almost 50% reduction in drilling time in South Texas' Eagle Ford Basin due to employment of big data collected from deployed sensors in the shale [13].

The upcoming chapters provide a thorough understanding of concepts related to various aspects of big data in exploration and drilling operations among others, including applications, benefits, characteristics, implementation, big data platforms, research, workflow, real-world implementations, successful implementation company traits, challenges and future scope. They provide a deep dive into big data which is otherwise lacking in currently available literature and research work, as noted by Vega-Gorgojo et al. [14].

2.3   RESERVOIR STUDIES

Reservoir studies, also known as reservoir engineering, refers to a branch of study/ working under petroleum industry that deals with the scientific principles and applications of hydrocarbon flow through porous subsurface medium during the drilling (development) and production operations from petroleum reservoirs in order to obtain a high economic recovery. The study consists of (but not limited to) subsurface geology, applied mathematics, technological integration processes and basic laws of physics & chemistry.

In order to understand reservoir study, it is important to first understand reservoirs. There is no simple definition of a petroleum reservoir, as it exhibits a set of different environmental conditions of rocks under the surface that lead to hydrocarbons reserves. Yet, in a simplified manner, a reservoir refers to a formation or subsurface pool of rocks (porous in nature) where the hydrocarbons have accumulated over the years due to the process of diagenesis (a process of rock compaction which changes the physical and chemical properties of the rock) of buried organic materials (like remains of dead organisms) under extreme high temperature and pressure [15].

These hydrocarbons usually form in the source rock and then migrate upwards to eventually be collected into small and connected pore spaces in the rock while being trapped within the reservoir area by adjacent and overlying impermeable layers of rock (also termed as a reservoir seal) after replacing the originally present water and can comprise a variety of particular mixtures (depending on the original biological material present, reservoir's temperature and pressure). They are then extracted from the trap (cumulation of source rock, hydrocarbons and seal) during the drilling and production operations [16,30].

According to petroleum geologists, the formation of the petroleum reservoirs or traps occur due to the following geological conditions [17]:

*   Rich source rocks that contain organic materials for hydrocarbon production
*   Sufficient heating of the source rock that liberates the crude oil
*   A reservoir-spaced cavity where liberated hydrocarbons can be accumulated
*   The cavity must be porous and permeable enough to store as well as transfer the hydrocarbons
*   It must also possess "impermeable" cap rock that prevents escape of hydrocarbons to the surface
*   Source rock, cavity and the cap rock must be aligned in a manner that allows the trapping of hydrocarbons

These attributes and conditions can be clearly seen in Figure 2.3 [18] that summarizes the formation of hydrocarbons under the subsurface along with portrayal of source rock, migration pathway, surrounding water, oil, gas and cap rock. Furthermore, as perceived from the conditions, it is obvious to note that sandstone and carbonate rocks (types of sedimentary rocks) showcase the majority of petroleum reserves in the world. Other than these, there are shale (clay-based rocks) reserves that also consist of petroleum reserves; however, sandstone reservoirs are of higher quality than carbonate and shale reservoirs in general [19].

Moreover, reservoirs have simple classification, namely, conventional and unconventional reservoirs. The conventional reservoirs refer to traditional reservoirs with low permeability, whereas unconventional reservoirs refer to reservoirs with low permeability and high porosity (keeping reservoir intact without the need for cap rock).

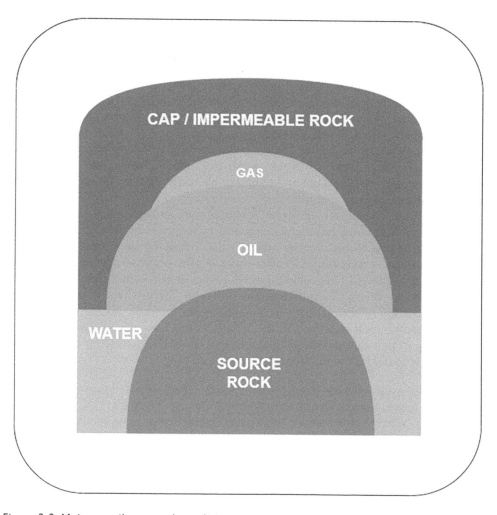

Figure 2.3 Major attributes and conditions pertaining to a reservoir.

Once the concept of a reservoir is apprehended properly, it is vital to note various aspects of reservoir engineering, such as Petrophysics, Exploration Geology/Seismology, Petroleum Economics, Production Operations and Process Engineering; all are summarized in Figure 2.4 [18,20]. These aspects will play a major role in applications and benefits of big data in the upcoming chapters.

Moving forward, one of the core functions of reservoir studies lies in the accurate estimation of the amount of petroleum reserves inside the reservoir for several reasons including comprehension of economic estimates of operations, legal finance reporting to regulatory bodies, approval(s) from regulatory bodies, negotiation of property sales & acquisitions, determination of market value, design of facilities, evaluation of profit and/or interest, planning & development of national energy policies, effective investment in the sector and reconcilement of any involved arbitration [21] (summarized in Figure 2.5). Other important functions of reservoir studies also include numerical

Figure 2.4 Major aspects of reservoir study.

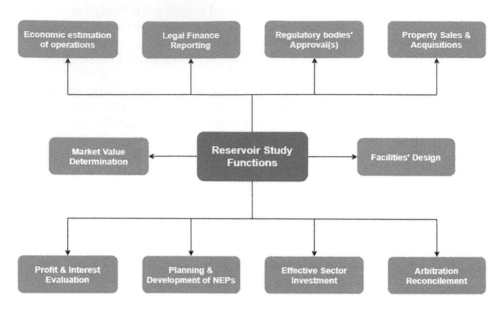

Figure 2.5 Major reservoir study functions.

reservoir modelling, production forecasting, well testing, well drilling and workover planning, economic modelling and PVT analysis of reservoir fluids.

As the major role of reservoir studies is the estimation of petroleum reserves, so is the role of reservoir engineers. This requires them to calculate two vital values, namely, OIP (oil in place, which is the amount of hydrocarbons filled inside the reservoir) and RF (recovery factor, which is the recoverable amount of petroleum initially in place that's usually expressed in terms of percentage) [22].

While the use of calculus and mathematics has deliberately been kept at minimum due to major focus on big data, a few concepts do require it, and thus a viable effort has been put into explaining it. For instance, both the terms, oil in place and recovery factor, are vital to understand, and thus their calculation method and related terms have been explained in the chapter. The value of OIP is calculated using the following formula:

$$OIP = V \cdot \Phi \cdot \left(1 - S_{wc}\right)$$

where

V = net bulk volume of the reservoir rock

Φ = porosity, or volume fraction of the rock which is porous

S_{wc} = irreducible water saturation, expressed as a pore volume fraction

On a side note, it is vital to understand from a theory perspective that the product of $V \cdot \Phi$ represents the pore volume, which is the amount of volume inside reservoir that fluids can take up. The products $V \cdot \Phi \cdot \left(1 - S_{wc}\right)$, on the other hand, represent the hydrocarbon pore volume, which is the volume inside reservoir that hydrocarbons (oil or gas or both) can take up [23].

The value of RF is usually simple and can be calculated using:

$$RF = \frac{EUR}{IGIP}$$

where
   EUR = Estimated Ultimate Recovery
   IGIP = Initial Gas In Place

On another side note, EUR refers to an approximation of petroleum quantity that can be potentially recovered or has already been while IGIP refers to the original approximate amount of gas inside the reservoir [24].

Furthermore, as discussed about the role of petroleum engineers in previous sections, reservoir studies require specialized reservoir engineers who play a central role in various aspects of reservoir operation and study, majorly including field development planning and recommending feasible & cost-effective schemes for reservoir depletion (such as waterflooding, chemical injection or gas injection for maximum petroleum recovery). Inclusively, their role includes reservoir evaluation, production forecasting, reserves estimation, field management, building numerical reservoir models and well testing & analysis as well [20]. Furthermore, due to legislative changes in several petroleum extensive countries, reservoir engineers also work on design and implementation of carbon sequestration projects that take care of environmental impact control through reduction of greenhouse emissions.

On a side note, on top of expertise and work pertaining to petroleum engineers (as mentioned in the previous sections), reservoir engineers have to perform several tasks (pertaining to aforementioned roles) which include:

- Estimation of the original oil in place (OIP)
- Evaluation of the hydrocarbon recovery factor (RF)
- Assignment of a timeline to the process of hydrocarbon recovery
- Collection, management and analysis of involved data
- Development and implementation of reservoir optimization techniques
- Undertaking of reservoir characterization
- Design and coordination of petrophysical studies
- Risk assessment of reservoir-related operations
- Prediction of recovery methods' (like EOR) performance
- Assessment of economics of major development programmes
- Analysis of pressure transients
- Development of effective monitoring and surveillance programmes

Now that the fundamentals have been cleared up, lets transition into the advent of big data into reservoir studies. Considering the fact that reservoir engineering involves mathematics and physics-based calculations for working of its several aspects, big data possesses massive potential for bringing benefits and applications like no other (discussed later in the upcoming chapters). This is the result of insights that it provides through an input of cumulation of different data sources, by passing through big data analytics (using various tools and technologies) as portrayed in Figure 2.6.

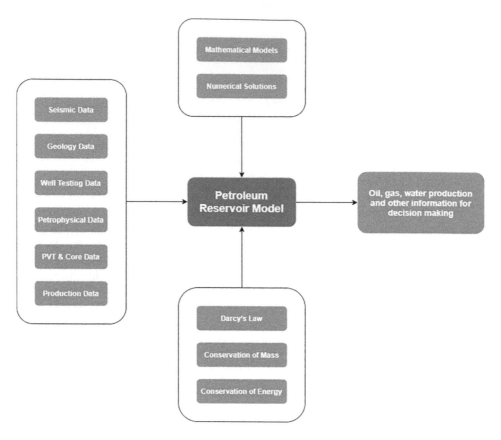

Figure 2.6 Big data insights in reservoir studies' modelling.

Big data solutions help in both collection and processing of data that petroleum companies require to make their reservoir production effective in every aspect (time, expenditure, efficiency, etc.). The data is collected through a network of downhole sensors (such as pressure, temperature and acoustic sensors) upon which big data analytics can be performed by companies to develop reservoir management applications. These applications ultimately provide timely as well as actionable information about changes in reservoir pressure, temperature, flow and acoustics, resulting in more control of companies over their operations along with increase in the profitability of reservoir. The intricate benefits and applications that broaden the big data's implementation mentioned above will be discussed in upcoming chapters.

2.4 PRODUCTION AND TRANSPORTATION

After the petroleum exploration has taken place, drilling process has been completed and reservoir modelling is successful, the next essential steps happen to be the actual production of hydrocarbons from the reservoir and its transportation via midstream operations to downstream operations' locations. The production of hydrocarbons is no easy feat (in terms of expertise, complexity of operation and level of execution-precision required) and involves several hazards such as reservoir leakage that make it risky affair. The same is true with transportation too, as it requires special expertise, solving of complex problems (such as cost minimization and route fixation) and is surrounded by hazards (especially petroleum leaks). This concerns everyone involved as the main goal here is to reduce the complexity of operations while minimizing the risks associated with it as much as possible.

In order to understand the role of big data in production and transportation, it is vital to first understand about both the operations briefly. Petroleum production, also termed as recovery of petroleum, refers to the set of operation that take place in order to extract the raw hydrocarbons from the underground reservoir. It consists of actual extraction and its maintenance. Intricately, it also includes the control of production & injection for meeting approved plans (of volumes and products' quality), the recording of all data (reservoir, wells and facilities) and monitoring of the equipment used across all operations for hassle-free workflow.

It itself has sub-phases for extraction, i.e. initial Build-Up period (where the production wells are progressively brought on flow with consistent increase in hydrocarbons' extraction), Plateau period (where a consistent and constant flow of hydrocarbons takes place) and Decline period (where the extraction rate or amount of hydrocarbons generates consistently slows down).

Moreover, production also comprises other utility systems that do not directly participate in the extraction process but aid it significantly, such as:

- power system (fuel gas and diesel)
- water and potable water treatment system
- chemicals and lubrication oils alarm
- fire protection and fire-fighting system
- instrument/utility air system
- shutdown system

On the other side, as most of the petroleum production in the world occurs at remote locations around the world when compared to where it will be consumed, transportation becomes a necessary part of the process (taking almost entirety of midstream segment) in order to facilitate the movement of produced raw hydrocarbons to processing plants or refineries. This facilitation happens via four major hereby mentioned options:

- **Pipeline**
  Pipelines happen to be the most widely utilized means of transportation for petroleum, majorly for moving crude oil from drilling site to refineries. They are preferred due to their less need for energy and relatively low carbon footprint.
- **Rail**
  As new oil reserves are found across the globe, rail transportation has picked up pace due to several reasons including relatively small capital costs and less construction period, making it a viable alternative to pipelines. However, reduced speed, higher energy consumption and grater carbon footprint happen to be its major drawbacks.

- **Ship**

  When land transportation is not viable (due to physical or economic hurdles), the transportation likely takes place through ships over water bodies. The fact that a single ship can accommodate petroleum of huge size (in factors of ten compared to rail tanks) at less than half the cost portrays an economically viable option for O&G companies. The drawbacks however are significant, in terms of speed and environmental impacts (such as leaks).

- **Truck**

  Transportation through trucks takes place in situations when a good amount of flexibility is required for the destination. As being the most limited form of transportation, it is usually used as a last step in transportation (such as moving of refined products from refineries to storage facilities).

Once the concepts of a transportation and production are comprehended properly, it is important to note various aspects of them as well, such as Petroleum Economics, PetroStatistics and governmental regulatory restrictions, all summarized in Figure 2.7 [18,20]. These aspects too will play a major role in applications and benefits of big data (discussed later in the upcoming chapters).

As discussed about the role of petroleum engineers earlier, production and transportation operations also require specialized petroleum engineers who play a central role in various aspects of operations (mentioned above) and studies, majorly including recommendation feasible & cost-effective schemes, field management, logistic handling and overseeing maintenance operations [20]. Furthermore, due to legislative changes in several petroleum extensive countries, alike reservoir engineers, engineers involved in production and transportation have to work on design and implementation of carbon impact reduction projects and hazards prevention projects as well (that take care of environmental impact control).

On a side note, on top of expertise and work pertaining to petroleum engineers (as mentioned in the previous sections), the engineers involved in production and transportation have to perform several tasks (pertaining to aforementioned roles) that briefly include:

- Determination of the rate of recovery for hydrocarbons (build-up, plateau and decline)
- Assignment of a timeline to the process of hydrocarbon recovery
- Collection, management and analysis of involved data
- Development and implementation of optimization techniques
- Coordination of petro-statistics aspects of operations
- Risk assessment of extraction and transportation operations
- Assessment of economics of major development programmes
- Development of effective monitoring and surveillance programmes

Once the fundamentals have been cleared up, again transitioning into the advent of big data into production and transportation operations, these operations involve complex statistical analysis for working of its key aspects (such as extraction and logistics); big

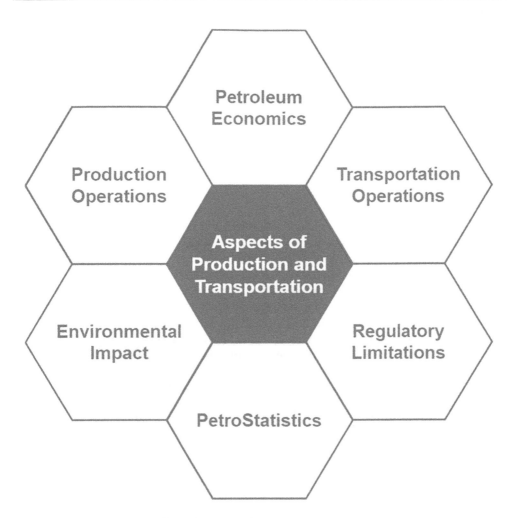

Figure 2.7 Major aspects of production and transportation.

data possesses massive potential for bringing benefits and applications like no other (discussed later in the upcoming chapters). This is the result of insights that it provides through an input of cumulation of different data sources (such as sensors attached to production/extraction gear and transportation vehicles or equipment), by passing through big data analytics (using various tools and technologies).

Big data solutions help in collection, processing and analysis of data that petroleum companies require to make their operations effective in every aspect (time, expenditure, efficiency, etc.). The data is collected through a network of sensors (such as pressure, temperature and acoustic sensors) in various areas (such as wells, work equipment, pipelines and storage facilities) upon which big data analytics can be performed by companies to develop strategic techniques of optimization.

2.5 PETROLEUM REFINERY

After the production and transportation operations have taken place, the next essential step happens to be the refining of the transported hydrocarbons from the reservoir. The refining of hydrocarbons is no easy feat (in terms of expertise, complexity of operations and level of execution-precision required) and involves several activities (such as operation optimization and cost minimization) and hazards (such as refinery leakage) that make it a complex and risky affair [25]. This concerns everyone involved as the main goal here is to reduce the complexity of operations while minimizing the risks associated with it as much as possible (while taking environmental aspects into consideration).

In order to understand the role of big data in petroleum refinery, it is vital to first understand about its operations briefly. A major part of entire downstream segment is the petrochemical refining of raw crude oil, which is a mixture of thousands of different hydrocarbons (where each component has its own weight, density, size, texture and importantly, boiling temperature). In a refinery, these components can be separated through various processes and methods (such as careful application of heat to capture various parts (called fractions) within certain boiling ranges, also termed as distillation).

The finished products from the refining operation(s) are categorized into three main types – Light Products (Liquid petroleum gas, Gasoline/Petrol and Naphtha), Medium Products (Kerosene, Jet/Aircraft fuels and Diesel fuel) and Heavy Products (Lubricating oils, Paraffin wax, Petroleum Coke and Asphalt/Tar). An elaborated comprehension of the refining of different components can be understood with the help of Figure 2.8, showing the fractional distillation process of filtering raw petroleum constituents [26,27].

Moreover, even after the filtration process is completed, the filtered products cannot be sold directly as they need further processing using both heat and pressure to improve qualities and meet market demand. This is one of the major reasons for the conversion of "unwanted heavy fuel oil into marketable gasoline and diesel", which is one of the major refinery operations [28].

Once the concepts of a refinery are comprehended properly, it is important to note various aspects of them as well, such as Petroleum Economics, Petrophysics and governmental regulatory restrictions, all summarized in Figure 2.9 [18,20]. These aspects will play a major role in applications and benefits of big data for the same in the upcoming chapters.

As discussed about the role of petroleum engineers in previous sections, petroleum refining operations also require specialized petroleum engineers who play a central role in various aspects of operations (mentioned above), majorly including identification of crude O&G products' key characteristics, recommendation of feasible & cost-effective operational schemes and overseeing of maintenance operations [20,28]. Furthermore, due to legislative changes in several petroleum extensive countries, alike reservoir engineers, engineers involved in refining operations have to work on design and implementation of carbon impact reduction projects and hazards prevention projects (or schemes) as well that take care of environmental impact control.

On a side note, on top of expertise and work pertaining to petroleum engineers (as mentioned in the previous sections), the engineers involved in refining have to perform several tasks (pertaining to the aforementioned roles) that briefly include:

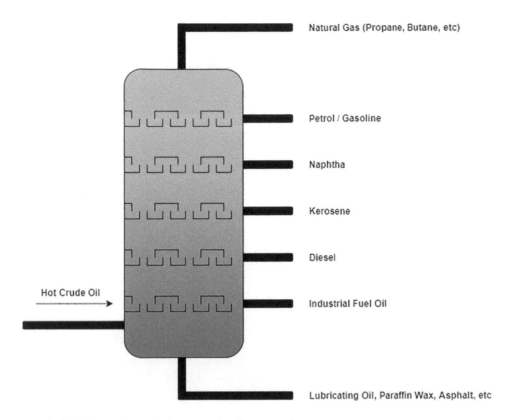

Natural Gas (Propane, Butane, etc)

Petrol / Gasoline

Naphtha

Kerosene

Diesel

Hot Crude Oil

Industrial Fuel Oil

Lubricating Oil, Paraffin Wax, Asphalt, etc

Figure 2.8 Refining through fraction distillation column.

- Demonstration of key refinery processes used for the conversion of raw hydrocarbons to useful final products
- Recognition of how refiners can alter the operation of a refinery
- Description of the major business processes for management of refinery
- Explanation of the key business drivers that impact refinery profitability
- Definition of environmental regulations' impact on refineries
- Understanding of major global refinery industry trends
- Assignment of a timeline to the refinery process
- Collection, management and analysis of involved data
- Development and implementation of optimization techniques
- Coordination of petrophysics aspects of operations
- Risk assessment of refining operations
- Development of effective monitoring and surveillance programmes

After the fundamentals have been cleared up, considering the fact that all refinery operations involve complex analysis for working of its key aspects (such as filtering and processing), big data possesses massive potential for bringing benefits and applications like no other (discussed later in the upcoming chapters). This is the result of insights that it provides through an input of cumulation of different data sources (such

Figure 2.9 Major aspects of petroleum refining.

as sensors attached to refinery gear), by passing through big data analytics (using various tools and technologies).

Big data solutions help in collection, processing and analysis of data that petroleum companies require to make their operations effective in every aspect (time, expenditure, efficiency, etc.). The data is collected through a network of sensors (such as pressure, temperature and acoustic sensors) in various areas (such as distillation column and work equipment) upon which big data analytics can be performed by companies to develop strategic techniques of optimization.

2.6 NATURAL GAS

Other than the crude oil gained from petroleum, a major product for petroleum is the gas refined from the raw hydrocarbons during the refining operations, which comprises methane, natural gas liquids (NGL, which are also hydrocarbon gas liquids) and non-hydrocarbon gases (such as carbon dioxide and water vapour). It has an absence of colour, odour & taste and is non-toxic in nature, and its generation of natural gas occurs below the earth surface under extreme conditions in a petroleum reservoir.

In a simplified manner (as mentioned previously), a petroleum reservoir refers to a formation or subsurface pool of rocks (porous in nature) where the hydrocarbons have accumulated over the years due to the process of diagenesis (a process of rock compaction which changes the physical and chemical properties of the rock) of buried organic materials (like remains of dead organisms) under extreme high temperature and pressure [29].

Interestingly, in some places, natural gas moves into large cracks or spaces between layers of the overlying rock, and the natural gas found in such formations is usually termed as conventional natural gas. While in other places, natural gas occurs in tiny pores or spaces (within some formations of shale, sandstone and other types of sedimentary rock), and this natural gas is usually referred as shale or tight gas and termed as unconventional natural gas. It also occurs with crude oil deposits, and this natural gas is termed as associated natural gas [31].

Furthermore, according to petroleum geologists, the formation of the petroleum reservoirs or traps occurs due to the following geological conditions [32]:

- Rich source rocks that contain organic materials for hydrocarbon production
- Sufficient heating of the source rock that liberates the crude oil
- A reservoir-spaced cavity where liberated hydrocarbons can be accumulated
- The cavity must be porous and permeable enough to store as well as transfer the hydrocarbons
- It must also possess "impermeable" cap rock that prevents escape of hydrocarbons to the surface
- Source rock, cavity and the cap rock must be aligned in a manner that allows the trapping of hydrocarbons

These attributes and conditions lead to generation of gas on top of heavy hydrocarbons, as clearly seen in Figure 2.3 [18], which summarizes the formation of hydrocarbons under the subsurface along with portrayal of source rock, migration pathway, surrounding water, oil, gas and cap rock.

Natural gas is used in various areas such as power generation, domestic chores, power engines, transportation, animal & fish feed, fertilizer creation and hydrogen fuel. The reach and applicability of natural gas makes it a vital part of petroleum industry, and hence, big data's implementation in the same has tremendous positive potential outcomes.

Moving forward, in order to understand the role of big data in natural gas, it is vital to first briefly understand about its refining process. A major part of entire downstream segment is the petrochemical refining of raw hydrocarbons which eventually give natural gas. In a refinery, the gas components (like liquid components) can be separated through various processes and methods (such as careful application of heat to capture various parts (called fractions) within certain boiling ranges, also termed as distillation). As stated earlier, the refining operation(s) are categorized into three main types – light, medium and heavy products. In the same, the lightest parts (such methane, propane and butane) are refined on the top of distillation column.

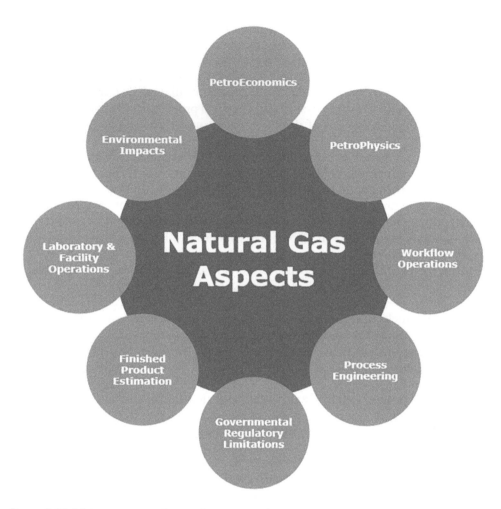

Figure 2.10 Major aspects of petroleum natural gas.

Once the concepts of natural gas and refining operations are comprehended properly, it is important to note various aspects of them as well, such as Petroleum Economics, Petrophysics and governmental regulatory restrictions, all summarized in Figure 2.10 [18,20].

After noting the aspects related to natural gas, it is essential to understand the six steps in the operations of producing natural gas, consisting of extraction (of petroleum), treatment (of natural gas), transportation (via pipelines, trucks, ships, etc.), storage (in specialized containers), distributions (among various channels and distributors) and decommissioning (of natural gas reservoirs), summarized in Figure 2.11 [33]. These operations are the backbone of natural gas production and thus are required to comprehend the entire process in detail.

As discussed about the role of petroleum engineers previously, natural gas workflow operations also require specialized petroleum engineers who play a central role in various aspects of operations (mentioned above), majorly including identification

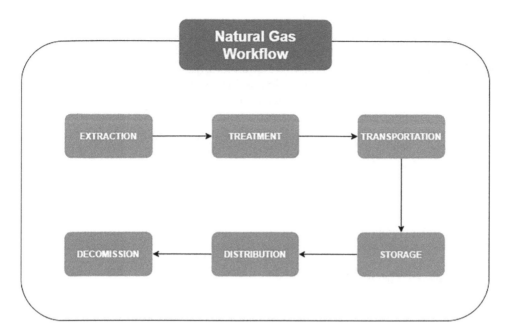

Figure 2.11 Petroleum natural gas operations workflow.

of recommendation of feasible & cost-effective operational schemes, risk assessment and mitigation and overseeing of maintenance operations [20,28]. Furthermore, due to legislative changes in several petroleum extensive countries, alike reservoir engineers, engineers involved in natural gas workflow operations have to work on design and implementation of carbon impact reduction projects and hazards prevention projects (or schemes) as well that take care of environmental impact control.

On a side note, on top of expertise and work pertaining to petroleum engineers, the involved engineers have to perform several tasks (pertaining to aforementioned roles) that briefly include:

- Demonstration of key extraction, refinery and maintenance processes used for the conversion of raw hydrocarbons to natural gas
- Description of the major business processes for management of natural gas pertaining operations
- Explanation of the key business drivers that impact natural gas' set of operations' profitability
- Definition of environmental regulations' impact on operations
- Assignment of a timeline to the natural gas generation processes
- Collection, management and analysis of involved data
- Development and implementation of optimization techniques
- Coordination of petrophysics aspects of operations
- Understanding of major global natural gas industry trends
- Development of effective monitoring and surveillance programmes
- Risk assessment of entire operation

Now that the fundamentals have been cleared up, transitioning again into the fact that all pertaining operations (including gas treatment operations) involve complex analysis for working of its key process aspects, big data possesses massive potential for bringing benefits and applications like no other. This is indeed the result of insights that it provides through an input of cumulation of different data sources (such as sensors attached to extraction, transportation and refinery gear), by passing through big data analytics (using various tools and technologies).

Big data solutions help in collection, processing and analysis of data that petroleum companies require to make their operations effective in every aspect (time, expenditure, efficiency, etc.). The data is collected through a network of sensors (such as pressure, temperature and acoustic sensors) in various areas (such as drilling wells, work equipment, pipelines, transportation vehicles and distillation column) upon which big data analytics can be performed by companies to develop strategic techniques of optimization (that could potentially save billions). It is also to be noted that the application of big data in natural gas helps in avoidance of several operational hazards (such as possibility of explosions, worker safety risks, greenhouse emissions and environmental effects).

2.7   HEALTH AND SAFETY

During the execution of all the operations in oil and gas industry, an essential step happens to be the health and safety of the petroleum professionals who interact with the industry in one form or the other. The activities pertaining to ensuring petroleum health and safety is no easy feat (in terms of expertise, complexity of involved operations and level of execution-precision required) and involves several aspects (mentioned later) and hazards (also discussed later) that make it a complex yet life-saving affair [25]. This concerns everyone involved as one of the main goals of petroleum industry is to ensure minimization of the operation associated risks and safety improvements as much as possible (while taking environmental aspects into consideration).

Moreover, in order to understand the role of big data in petroleum health and safety, it is vital to first understand about health and safety's importance in the industry. Petroleum prices may fluctuate and so does the budget constraints & cut, but the one that stays the same is the protection of industry's most valuable asset – employees. The petroleum professionals drive the entire industry and their protection drive industry's profits as well, which makes the implementation of Health and Safety operations very important.

Therefore, the benefits and applications that big data possesses to cater to the improvement of health and safety operations tend to make it a top priority for oil and gas companies. A successful implementation of big data in health and safety operations helps in the prevention and control of various safety, injury, illness and health hazards summarized in Figure 2.12.

Petroleum professionals are subject to risks on a daily basis from all petroleum operations. This level of work-related risks is usually unseen in most other industries, and thus, it pushes O&G companies to make the workspace safer to work in. And this portrays a major upside to the implementation of big data in health and safety operations.

Once the concepts of health and safety are comprehended properly, it is important to note various aspects of them as well, such as monitoring and surveillance among others summarized in Figure 2.13. These aspects will play a major role in applications and benefits of big data for the same in the upcoming chapters.

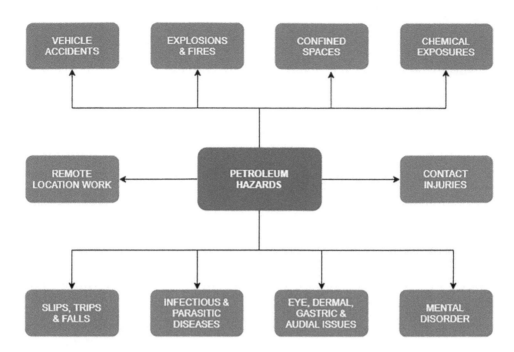

Figure 2.12 Major hazards of petroleum industry.

Petroleum health and safety operations also require specialized petroleum engineers who play a central role in various aspects of aforementioned operations, majorly including overseeing of maintenance operations, safety regulations monitoring & enforcement, and assessment of health & safety equipment & processes. Furthermore, due to legislative changes in several petroleum extensive countries, alike reservoir engineers, engineers involved in health and safety operations play a role in design and implementation of carbon impact reduction projects and hazards prevention projects (or schemes) as well that take care of environmental impact control.

On a side note, on top of expertise and work pertaining to petroleum engineers, the engineers involved in health and safety have to perform several tasks (pertaining to aforementioned roles) that briefly include:

- Demonstration of key processes used for the health and safety of petroleum professionals
- Recognition of different processes that can alter the health and safety operations
- Description of the major business processes for management of health and safety operations
- Definition of environmental regulations' impact on petroleum operations
- Assignment of a workflow to the health and safety processes
- Collection, management and analysis of involved data
- Development and implementation of optimization techniques
- Risk assessment of all petroleum operations
- Development of effective monitoring and surveillance programmes

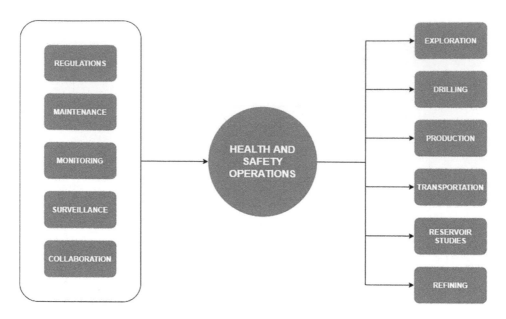

Figure 2.13 Major aspects of petroleum health and safety.

Moreover, there are various safety guidelines that ensure enhanced safety and hazard prevention for O&G companies and worker [34], which briefly include:

• Invest in safety programmes, equipment and processes
• Monitor workers' mental health on consistently
• Possess never-stop-learning mentality while being aware of all hazards
• Install in-vehicle monitoring systems for transportation
• Ensure work-site familiarity for better hazardous situation response(s)
• Collaborate effectively with local emergency response communities
• Promote healthy communication within the workforce
• Reassess safety signage with each project shift
• Ensure timely and consistent equipment maintenance
• Provide clear visual communication throughout work-site(s)
• Implement a 5S (Sort, Store, Shine, Standardize and Sustain) system
• Maintain consistent "housekeeping" of the work-site

Considering the fact that all petroleum operations have health and safety associated with it and involve complex analysis for working of its key aspects, big data possesses massive potential for bringing benefits and applications unparalleled to others. This is the result of the insights that it provides through an input of cumulation of different data sources (such as sensors attached to equipment gear), by passing through big data analytics (using various tools and technologies).

Big data solutions help in collection, processing and analysis of data that petroleum companies require to make their operations safer and greener in every aspect (health, working environment and associated risks among others). The data is collected through a network of sensors (such as pressure, temperature, acoustic, moisture and fault detection sensors) in various areas upon which big data analytics can be performed by companies to develop strategic techniques of optimization.

2.8   FINANCE AND FINANCIAL MARKETS

Oil and Gas industry is one of the biggest sectors or portions of all industries in the world in terms of monetary value (as high as $2 trillion per annum in 2021 [35]). Due to its inherent nature of providing the most crucial economic framework – Oil to the world, Petroleum industry stands apart as one of the most significant, widespread and impactful industries of them all, especially for countries including the United States, Russia, Canada, Saudi Arabia and China. The reach of the industry is unfathomable, as it employs hundreds of thousands (directly) and millions (indirectly) of people around the globe, generates trillions of dollars of revenue and serves as the backbone of any nation's economy (due to its significant contribution to GDP – Gross Domain Product).

The industry today quenches 57% of global energy consumption, while roughly accounting for the same percentage of global CO_2 emissions. It also accounts for 1/10th of total global stock market capital, 15% of global exports and 25 of OPEC's (Organization of the Petroleum Exporting Countries) GDP. The industry also requires intense capital, expensive equipment and highly skilled labour [36] and supports a huge workforce of skilled and specialized professionals (10 million in the US alone). These stats can be comprehended in Figure 2.14.

Considering the importance of oil and gas in financial markets, it is also vital to note that "financing" of the industry happens to play a major role in each and every aspect of operations. Likewise, it is also important to understand the outcomes and

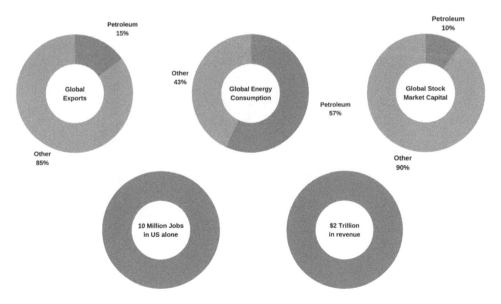

Figure 2.14 Petroleum industry statistics.

upsides of application of big data to the complex intricacies of finance and financial markets in the petroleum industry.

A universal aspect of any industry happens to be the proper management of expenditure for effective working of its activities while maximizing the profit margins. Therefore, cost management is a vital aspect to understand everywhere. In lieu of this, the major factors influencing the cost management in the industry can be understood via Figure 2.15. The O&G companies tend to majorly balance out these factors in order to maximize output and profits.

Talking about the financial aspects of the upstream segment, it is majorly categorized as high risk–high return segment due to enormous risks and heavy investment ($5–$20 million per site for exploration alone) associated with it along with substantial profits in case everything goes well. It is also highly regulated by nations' policies due to several factors mentioned before, which significantly impact the entire supply chain.

Figure 2.15 Cost management in petroleum industry.

Moreover, the segment is also heavily impacted by global politics (or state affairs) including political instabilities (like war, and/or international conflicts), laws & regulations (like environmental restrictions & social programmes), state-imposed price controls, tax regimes and respect for contracts. Another major aspect of upstream happens to be the uncertainty of production and its outcome, as several factors (like supply & demand, economic growth, recessions, crude production quotas, seasonal weather patterns and severe weather events' disruptions) play a major role in financial success (or profitability) of upstream operations.

The midstream segment comprises high regulations (due to catastrophic environmental consequences in case of mishaps) and low capital risk (due to its inherent non-risk profile of nature of operations and relatively less complexity of assets compared to other segments). It is to be noted that the segment's financial success is dependent on some key factors including the success of upstream firms on continuous delivery of reserves, refinery margins that encourage refined product production, health of every segment's consumer markets (due to its positioning), petroleum price levels (that impact the profitability of operations) and finally, the political sentiment of states for pipeline expansion(s) (due to "not in my backyard"-based perspective hurdles).

The business aspect of downstream segment, on the other hand, resides in the margins and thus is also called a margin business. A margin business is characterized by its working where the profits are made by the amount of margin present between the price of finished marketable products sold to consumers (here, petrochemical products) and the price of raw materials (here, crude oil brought in from upstream or midstream segment).

The profitability of downstream segment is also an interesting aspect to look at. While most integrated oil companies (IOCs) and other segment companies get hurt by lower oil prices (say, due to oversupply of petroleum), the downstream segment and its companies benefit from it substantially. This happens due to the fact that when petroleum prices fall sharply, petrochemical products lag the petroleum prices, and as a result, the refining margins grow (making substantial profits from the same). The converse is true too, as an increase in petroleum prices may result in declined profit margins.

Furthermore, the level of margins or profits for downstream companies also varies and can be affected by seasons (petroleum production has seasonality as well), parallel crude prices (of course), supply & demand (at that point of time) and production cuts (by institutions like OPEC). All these factors, when aligned properly, have potential to inherently change the amount of profitability by a factor of 5 to 6.

Once the aforementioned concepts are comprehended properly, it is important to note various aspects of them as well, such as petroleum asset management and financial modelling, all summarized in Figure 2.16. These aspects will play a major role in applications and benefits of big data for the same in the upcoming chapters.

Operations pertaining to petroleum finance and financial markets also require specialized petroleum engineers who play a central role in various aspects of operations, majorly including recommendation of feasible & cost-effective operational schemes and overseeing of financial modelling. Furthermore, due to legislative changes in several petroleum extensive countries, engineers involved in operations pertaining to finance and financial markets have to pursue their work while keeping the carbon impact reduction, hazards prevention as well as environmental impact control into consideration as well (alike other major operation engineers).

Figure 2.16 Major aspects of petroleum finance and financial markets.

On a side note, on top of expertise and work pertaining to petroleum engineers, the engineers involved in finance and financial markets have to perform several tasks (pertaining to aforementioned roles) that briefly include:

- Collection, management and analysis of involved financial data
- Description of the major business processes for financial modelling
- Rigorous analysis and execution of asset management
- Explanation of the key business drivers that impact profitability
- Understanding of major global industry trends
- Keep a check on global O&G financial markets
- Assignment of a timeline to the various processes along with profitability
- Development and implementation of optimization techniques
- Risk assessment of various industry operations

Now that the fundamentals, aspects and roles have been cleared up, it is important to understand the advent of big data into petroleum finance and financial markets. Considering the fact that all associated operations involve complex analysis of some form (such as financial modelling, distribution management and financial market analysis), big data possesses massive potential for bringing splendid benefits and applications. This is the result of insights that it provides through an input of cumulation of different data sources (such as financial statements and time series data), by passing through big data analytics (using various tools, platforms and technologies).

Big data solutions help in collection, processing and analysis of data that petroleum companies require to make their operations effective in every aspect (time, expenditure, efficiency and especially financially). The data is collected through a set of aforementioned channels in various areas (such as financial aspects of operations and financial markets) upon which big data analytics can be performed by companies to develop strategic techniques of optimization that could potentially save billions. These applications ultimately provide timely as well as actionable information about changes in the aforementioned areas, resulting in increase in their operational profitability. The intricate benefits and applications that broaden the big data's implementation mentioned above will be discussed in the upcoming chapters.

REFERENCES

1. Easley, S. A. (2021). *What Is Geoscience, Exactly?* Southern New Hampshire University. https://www.snhu.edu/about-us/newsroom/2018/08/what-is-geoscience.
2. U.S. Geological Survey. What Is Geoscience? Youth and Education in Science. https://www.usgs.gov/science-support/osqi/youth-education-science/what-geoscience.
3. Texas Geosciences. Why Geosciences? https://www.jsg.utexas.edu/education/undergraduate/why-geosciences/.
4. Yan, G., Xue, Q., Xiao, K., Jianping, C., Miao, J., & Yu, H. (2015). An analysis of major problems in geological survey big data. *Geological Bulletin of China*, 34(7), 1273–1279. https://www.researchgate.net/publication/283474378_An_analysis_of_major_problems_in_geological_survey_big_data.
5. Bagherian, N. (2015). Why the Oil and Gas Industry Needs Data Democratization. https://datafloq.com/read/Oil-Gas-Industry-Needs-Data-Democratization/1731.
6. DiChristopher, T. (2015). Oil Firms are Swimming in Data They Don't Use. https://www.cnbc.com/2015/03/05/us-energy-industry-collects-a-lot-of-operational-data-but-doesnt-use-it.html.
7. McKinsey Global Institute (2015). The Internet of Things: Mapping the Value Beyond the Hype. http://globaltrends.thedialogue.org/publication/the-internet-of-things-mapping-the-value-beyond-the-hype/.
8. Dyson, R. (2016). The Oil and Gas Industry Needs to Think Big with Big Data. https://www.worldoil.com/sponsored-content/digital-transformation/2016/17/the-oil-and-gas-industry-needs-to-think-big-with-big-data.
9. Martin, R. (2015). Big Data Will Keep the Shale Boom Rolling. *MIT Technology Review*. https://www.technologyreview.com/2015/06/02/11273/big-data-will-keep-the-shale-boom-rolling/.
10. Baaziz, A., & Quoniam, L. (2014). How to Use Big Data technologies to optimize operations in Upstream Petroleum Industry. *The 21st World Petroleum Congress*. https://arxiv.org/ftp/arxiv/papers/1412/1412.0755.pdf.

11. Soofi, A., & Perez, E. (2014). Drilling for New Business Value: How Innovative Oil and Gas Companies are Using Big Data to Outmaneuver the Competition. https://cloudblogs. microsoft.com/industry-blog/uncategorized/2015/10/18/drilling-new-business-value-innovative-oil-gas-companies-using-big-data-maneuver-competition/.

12. Sukapradja, A., Clark, J., Hermawan, H., & Tjiptowiyono, S. (2017). Sisi Nubi Dashboard: Implementation of Business Intelligence in Reservoir Modelling & Synthesis: Managing Big Data and Streamline the Decision Making Process. Paper presented at the *SPE/IATMI Asia Pacific Oil & Gas Conference and Exhibition*, Jakarta, Indonesia, October 2017. https://onepetro.org/SPEAPOG/proceedings-abstract/17APOG/2-17APOG/D021S012R002/194087.

13. Boichenko, T. (2021). Big Data and Predictive Analytics in Oil and Gas. https://www.n-ix. com/big-data-predictive-analytics-oil-and-gas/.

14. Vega-Gorgojo, G., Fjellheim, R., Roman, D., Akerkar, R., & Waaler, A. (2016). Big Data in the Oil & Gas Upstream Industry – A Case Study on the Norwegian Continental Shelf. https://www.researchgate.net/publication/305416575_Big_Data_in_the_Oil_Gas_Upstream_Industry_-_A_Case_Study_on_the_Norwegian_Continental_Shelf.

15. Hanania, J., Le, C., Meyer, R., Sheardown, A., Stenhouse, K., & Donev, J. (2019). *Energy Education – Oil and Gas Reservoir*. University of Calgary. https://energyeducation.ca/encyclopedia/Oil_and_gas_reservoir.

16. Wheaton, R. (2016b). Basic Rock and Fluid Properties. Fundamentals of Applied Reservoir Engineering, 5–57. https://doi.org/10.1016/b978-0-08-101019-8.00002-8.

17. Selley, R.C., 1998. *Elements of Petroleum Geology*. Academic Press, San Diego, CA. https://www.sciencedirect.com/book/9780123860316/elements-of-petroleum-geology.

18. Okotie, S., & Ikporo, B. (2018b). Introduction. Reservoir Engineering, 1–73. https://doi. org/10.1007/978-3-030-02393-5_1.

19. Jamshidnezhad, M. (2015). Introduction. Experimental Design in Petroleum Reservoir Studies, 1–8. https://doi.org/10.1016/b978-0-12-803070-7.00001-6.

20. Wheaton, R. (2016a). Introduction. Fundamentals of Applied Reservoir Engineering, 1–4. https://doi.org/10.1016/b978-0-08-101019-8.00001-6.

21. Okotie, S., & Ikporo, B. (2018a). Resources and Reserves. Reservoir Engineering, 75–86. https://doi.org/10.1007/978-3-030-02393-5_2.

22. Schlumberger Oilfield Glossary. https://glossary.oilfield.slb.com/en/Terms/r/recovery_factor.aspx.

23. Dake, L. P. (1983). *Fundamentals of Reservoir Engineering*. Developments in Petroleum Science, Elsevier Science, Shell Learning and Development. https://www.elsevier.com/books/fundamentals-of-reservoir-engineering/dake/978-0-444-41830-2.

24. Chen, J. (2020). Estimated Ultimate Recovery (EUR), Investopedia. https://www.investopedia.com/terms/e/estimated-ultimate-recovery.asp.

25. Shah, N. K., Li, Z., & Ierapetritou, M. G. (2011). Petroleum refining operations: Key issues, advances, and opportunities. *Industrial & Engineering Chemistry Research*, 50(3), 1161–1170. https://doi.org/10.1021/ie1010004.

26. University of Calgary. Fractional distillation. Energy Education. https://energyeducation. ca/encyclopedia/Fractional_distillation.

27. Freudenrich, C. How Oil Refining Works. Howstuffworks. https://science.howstuffworks. com/environmental/energy/oil-refining4.htm.

28. EKT Interactive (2020). What Is Refining? https://ektinteractive.com/refining/.

29. Anderson, R. N. (2017). "Petroleum Analytics Learning Machine" for Optimizing the Internet of Things of Today's Digital Oil Field-to-Refinery Petroleum System. *2017 IEEE International Conference on Big Data (Big Data)*, 4542–4545. https://doi.org/10.1109/bigdata.2017.8258496.

30. Mohammadpoor, M., & Torabi, F. (2020). Big Data analytics in oil and gas industry: An emerging trend. *Petroleum*, 6(4), 321–328. https://doi.org/10.1016/j.petlm.2018.11.001.

31. U.S. Energy Information Association. Basics. Natural Gas Explained. https://www.eia.gov/energyexplained/natural-gas/.

32. Fidelis, E. W., & Effiong, A. (2019). Utilization of Big Data Analytics for Effective Refinery Operations. *NAPE Annual International Conference 2019*, 29(1), 113–120. https://nape.org.ng/wp-content/uploads/2020/09/113-120.-Utilization-of-Big-Data-Analysis.pdf.

33. Eni-Scoula. Natural gas. *Energy*, 1–14. http://www.eniscuola.net/wp-content/uploads/2011/02/pdf_natural_gas1.pdf.

34. Rzepecki, K. (2018). *Ten Safety Tips for Oil and Gas Industry Workers*. SME Society of Manufacturing Engineers. https://www.sme.org/technologies/articles/2018/november/ten-safety-tips-for-oil-and-gas-industry-workers/.

35. IBISWorld (2021). Global Biggest Industries by Revenue in 2021. https://www.ibisworld.com/global/industry-trends/biggest-industries-by-revenue/.

36. Kalyani, D. (2018). Oil Drilling & Gas Extraction in the US, IBISWorld Industry Report, 2018.

Chapter 3

Big Data's 6Vs

3.1 INTRODUCTION

Big data, in simple terms, can be attributed to a set of processes such as data acquisition, data storage, search, analysis and visualization, whose cumulation can result into deep understanding of data at hand in order to make meaningful and educated decisions in the real world. For instance, although Trump is not tech-savvy, a big data company (Cambridge Analytica) ensured that he won the elections with the help of big data. Alexander Nix, CEO of Cambridge Analytica, stated that virtually every message that Trump broadcasted was driven by big data [1].

Moreover, to understand it intricately, we need to define Vs associated with it, i.e., volume, variety, velocity, value, veracity and variability [1,2]. Volume is one of the straightforward aspects of big data and refers to data amount or size along with processing capacity of the same. Volume of big data has been on the rise for years now and is on the verge of exponential explosion in the next decade.

Variety on the other hand refers to the diversity in various types of data formats in which big data exists and is majorly divided into structured, unstructured and semi-structured data [3] along with a combination of two or all (although it has been reported that almost 90% of all data generated is unstructured [4]). Major data formats in the world today include audio, images, video, documents, emails, text files and graphical data among others.

Velocity of big data is the methodology of handling data generated at increasingly accelerated rates and transferring the same across devices, platforms or cloud services, and includes near time, real time, batches and streams. It also deals with time and latency aspects of data manipulation and management. Moreover, as the flow of data is often massive and consistent, it requires devices and platforms to possess capabilities to not only handle massive volumes, but deal with its stream in real-time as well.

Value refers to the use cases or benefits of using big data for creating value for the field where it is applied. It also involves understanding the potential to create revenue or uncover opportunities through big data. It simply outlines the results that the applications of big data cater.

Variability simply is change in big data and its model(s) for feasibility and includes management and contextualization of data in a way that provides "structure" to the implementation, even in unpredictable and variable data environments. Veracity refers to the level/quality of the big data being used by the companies, while including the

DOI: 10.1201/9781003185710-3

ability to leverage the potential of variable data sources and types (both structured and unstructured) to best use.

The summarized information about all the six Vs can be comprehended with the help of condensed information in Figure 3.1. As perceived from the Vs, it can be noted that the main goal of big data is to equip entities using it to process and analyse (using various big data tools) the complex data with feasible amount of time and resources, which would otherwise be impossible to accomplish with traditional methods.

This chapter enumerates the 6Vs of big data in each operation, starting with Geoscience (Section 3.2), Exploration and Drilling (Section 3.3), Reservoir Studies (Section 3.4), Production and Transportation (Section 3.5), Petroleum Refinery (Section 3.6), Natural Gas (Section 3.7), Health and Safety (Section 3.8) and Finance and Financial Markets (Section 3.9) followed by a conclusion (Section 3.10). Now, let us understand the 6Vs associated with various operations in the petroleum industry.

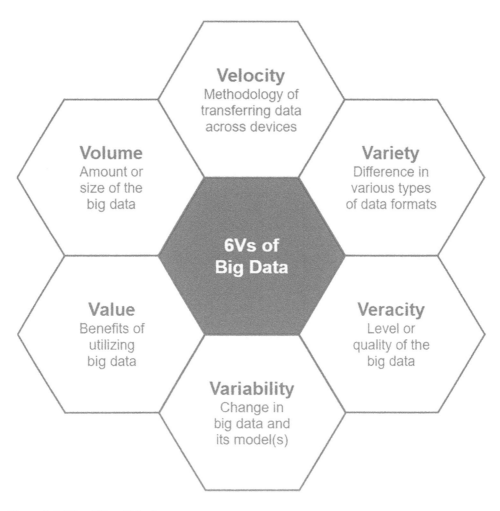

Figure 3.1 The 6Vs of big data.

3.2   BIG DATA'S 6VS IN GEOSCIENCE

All the aspects or 6Vs (Volume, Variety, Velocity, Veracity, Variability and Value) characteristics evidently apply to big data in geoscience as well, and the attributes related to them are hereby mentioned:

Volume:

- Earthen sensory data
- Automated SCADA
- Archived company data and processing

Variety:

- Structured (sensory data)
- Unstructured (images, maps, audio, video)
- Semi-structured (operational data analysis reports)

Value:

- Reduced safety hazards & environmental impacts
- Optimum and enhanced operations

Velocity:

- Real-time data streaming of data from activity-specific sensors
- Equipment-based varied data stream for big data analytics

Veracity:

- Activity-specific data cleaning and segregation
- Amalgamation of all data sources for a quality-rich dataset

Variability:

- Change in sensory big data
- Change in tools or technologies used for analysing the big data
- Usage of various machine learning models for analysis

3.3   BIG DATA'S 6VS IN EXPLORATION & DRILLING

All the 6V's aspects for big data in exploration and drilling along with its attributes are hereby mentioned:

Volume:

- Seismic data and processing
- Drilling rig data and processing
- Archived company data and processing

Variety:

- Structured (well logs and sensory data)
- Unstructured (images, maps, audio, video)
- Semi-structured (processed data analysis and sensory drilling reports)

Value:

- Optimum recovery of hydrocarbons for upcoming streams
- Efficient and effective raw material output for maximum profits
- Reduced safety hazards and environmental impacts
- Reduction in invisible non-productive time (INPT)
- Visually aesthetic portrayal of data at hand

Velocity:

- Exploration data constant transmission for n-dimensional imaging
- Real-time data streaming from well and drilling equipment sensors
- Real-time stream of LWD and MWD data
- Wide Azimuth data acquisition

Veracity:

- Seismic and drilling data cleaning and segregation
- Amalgamation of all data sources for a quality-rich dataset
- Combination of selected sources data for specific analysis

Variability:

- Change in exploration and drilling operations' big data itself
- Usage of various machine learning models for analysis
- Change in tools or technologies used for analysing the big data

3.4 BIG DATA'S 6VS IN RESERVOIR STUDIES

All the 6V's aspects for big data in reservoir studies along with its attributes are hereby mentioned:

Volume:

- Reservoir seismic data and processing
- Local-based external data acquisition
- Archived company data and processing

Variety:

- Structured (seismic data and sensory data)
- Unstructured (images, maps, audio, video)
- Semi-structured (processed data analysis reports)

Value:

- Reduced safety hazards & environmental impacts once development begins
- Optimum and enhanced recovery of hydrocarbons
- Visually aesthetic n-dimensional portrayal of data at hand
- Economic viability determination

Velocity:

- Real-time data streaming from seismic sensors for analysis
- Exploration data constant transmission for n-dimensional imaging
- Location-based varied data stream for big data analytics

Veracity:

- Seismic data cleaning and segregation
- Amalgamation of all data sources for a quality-rich dataset
- Combination of selected geological locations' data for specific analysis

Variability:

- Change in reservoirs' seismic data big data (via updated tools or process)
- Simulation variation due to change in visualization parameters
- Usage of various machine learning models for analysis
- Change in tools or technologies used for analysing the big data

3.5 BIG DATA'S 6VS IN PRODUCTION AND TRANSPORTATION

All the 6V's aspects for big data in production and transportation along with its attributes are hereby mentioned:

Volume:

- Production operations' sensor data
- Transportation operations' sensor data
- Automated SCADA
- Archived company data and processing

Variety:

- Structured (sensory data)
- Unstructured (images, maps, audio, video)
- Semi-structured (operational data analysis reports)

Value:

- Reduced safety hazards & environmental impacts
- Optimum and enhanced recovery of hydrocarbons

- Economic viability determination
- Operation-based optimization

Velocity:

- Real-time data streaming of data from operation-specific sensors
- Location-based varied data stream for big data analytics
- Evaluation-based varied data stream (flow, pressure, temperature, etc.) for big data analytics

Veracity:

- Operation-specific data cleaning and segregation
- Amalgamation of all data sources for a quality-rich dataset
- Combination of selected geological locations' data for specific analysis

Variability:

- Change in sensory big data (via updated tools, processes or real-world/external reasons)
- Change in tools or technologies used for analysing the big data
- Simulation variation due to change in visualization parameters
- Usage of various machine learning models for analysis

3.6 BIG DATA'S 6VS IN PETROLEUM REFINERY

All the 6V's aspects for big data in petroleum refinery along with its attributes are hereby mentioned:

Volume:

- Refinery operations' sensor data
- Automated SCADA
- Archived company data and processing

Variety:

- Structured (sensory data)
- Unstructured (images, maps, audio and video)
- Semi-structured (operational data analysis reports)

Value:

- Reduced safety hazards & environmental impacts
- Optimum and enhanced refining operations
- Economic viability determination and operational optimization

Velocity:

- Real-time data streaming of data from activity-specific sensors
- Equipment-based varied data stream for big data analytics
- Evaluation-based varied data stream (flow, pressure, temperature, etc.) for big data analytics

Veracity:

- Activity-specific data cleaning and segregation
- Amalgamation of all data sources for a quality-rich dataset

Variability:

- Change in sensory big data (via updated tools, processes or real-world/external reasons)
- Change in tools or technologies used for analysing the big data
- Simulation variation due to change in visualization parameters
- Usage of various machine learning models for analysis

3.7 BIG DATA'S 6VS IN PETROLEUM NATURAL GAS

All the 6V's aspects for big data in petroleum natural gas along with its attributes are hereby mentioned:

Volume:

- Extraction operations' sensor data
- Treatment operations' sensor data
- Storage and transportation operations' sensor data
- Distribution operations' sensor data
- Automated SCADA
- Archived company data and processing

Variety:

- Structured (sensory data)
- Unstructured (images, maps, audio and video)
- Semi-structured (operational data analysis reports)

Velocity:

- Real-time data streaming of data from activity-specific sensors
- Equipment-based varied data stream for big data analytics
- Evaluation-based varied data stream (flow, pressure, temperature, etc.) for big data analytics

Value:

- Reduced safety hazards & environmental impacts
- Optimum and enhanced natural gas generation operations
- Economic viability determination and operational optimization

Veracity:

- Activity-specific data cleaning and segregation
- Amalgamation of all data sources for a quality-rich dataset

Variability:

- Change in sensory big data (via updated tools, processes or real-world/external reasons)
- Change in tools or technologies used for analysing the big data
- Simulation variation due to change in visualization parameters
- Usage of various machine learning models for analysis

3.8   BIG DATA'S 6VS IN PETROLEUM HEALTH AND SAFETY

All the 6V's aspects for big data in petroleum health and safety along with its attributes are hereby mentioned:

Volume:

- Exploration operations' sensor data
- Drilling operations' sensor data
- Reservoir operations' sensor data
- Production operations' sensor data
- Transportation operations' sensor data
- Refining operations' sensor data
- Automated SCADA
- Archived company data and processing

Variety:

- Structured (sensory data)
- Unstructured (images, maps, audio, video)
- Semi-structured (operational data analysis reports)

Value:

- Reduced safety hazards & environmental impacts
- Optimum and enhanced operations
- Economic viability determination and operational optimization

Velocity:

- Real-time data streaming of data from activity-specific sensors
- Equipment-based varied data stream for big data analytics
- Evaluation-based varied data stream (flow, pressure, temperature, etc.) for big data analytics

Veracity:

- Activity-specific data cleaning and segregation
- Amalgamation of all data sources for a quality-rich dataset

Variability:

- Change in sensory big data (via updated tools, processes or real-world/external reasons)
- Change in tools or technologies used for analysing the big data
- Simulation variation due to change in visualization parameters
- Usage of various machine learning models for analysis

3.9   BIG DATA'S 6VS IN PETROLEUM FINANCE AND FINANCIAL MARKETS

All the 6V's aspects for big data in petroleum finance and financial markets along with its attributes are hereby mentioned:

Volume:

- Time-series financial market data
- Finance-based statement data
- Archived company data

Variety:

- Structured
- Unstructured
- Semi-structured

Value:

- Optimum and enhanced industry operations
- Economic viability determination and operational optimization
- Increased operational profitability

Velocity:

- Real-time data streaming of time series data
- Source-based varied data stream for big data analytics
- Evaluation-based varied data stream (flow, pressure, temperature, etc.) for big data analytics–based economic viability comprehension

Veracity:

- Activity-specific data cleaning and segregation
- Amalgamation of all data sources for a quality-rich dataset

Variability:

- Change in time series and operational finance big data (via updated tools, processes or real-world/external reasons)
- Change in tools or technologies used for analysing the big data
- Usage of various machine learning models for analysis

3.10 CONCLUSION

Considering the fact that petroleum operations are monitored (as surveillance and data gathering has been on rise) and all the previously mentioned 6Vs for each operation are properly comprehended, big data–employed analytics simply allows for a deeper look into data that is otherwise meaningless. Exploiting the best of the capabilities of big data and its 6Vs in the industry, O&G companies can see drastic improvements and massive upsides due to big data implementation in all their operations.

In a nutshell, these aspects or characteristics of big data 6Vs, employed onto aforementioned petroleum operations, result in various benefits (discussed in the upcoming chapter) that make it a worthwhile strategy to approach.

REFERENCES

1. Schaafsma, S. Big Data: The 6 Vs You Need to Look at for Important Insights. *Motivaction*. https://www.motivaction.nl/en/news/blog/big-data-the-6-vs-you-need-to-look-at-for-important-insights.
2. Jain, A. (2016). The 5 V's of Big Data, Watson Health Perspectives, IBM. https://www.ibm.com/blogs/watson-health/the-5-vs-of-big-data/.
3. Sumbal, M. S., Tsui, E., & See-to, E. W. (2017). Interrelationship between big data and knowledge management: An exploratory study in the oil and gas sector. *Journal of Knowledge Management*, 21(1), 180–196. https://doi.org/10.1108/jkm-07-2016-0262.
4. Ishwarappa, J., & Anuradha, J. (2015). A brief introduction on big data 5Vs characteristics and hadoop technology. *Procedia Computer Science*, 48, 319–324. https://doi.org/10.1016/j.procs.2015.04.188.

Chapter 4

Benefits of Big Data

4.1 INTRODUCTION

Big data refers to a set of processes such as data acquisition, data storage, search, analysis and visualization, whose cumulation can result in deep understanding of data at hand in order to make meaningful and educated decisions in the real world. These results pose massive benefits on the application field that are the reason for the boom in big data implementation.

Major benefits of big data implementation include massive computational upsides, efficiency improvements and data handling capabilities. The deeper dive into all the benefits will begin with the next section. This chapter outlines the industry-wide benefits (Section 4.2), benefits of big data in Geoscience (Section 4.3), Exploration and Drilling (Section 4.4), Reservoir Studies (Section 4.5), Production and Transportation (Section 4.6), Petroleum Refinery (Section 4.7), Natural Gas (Section 4.8), Health and Safety (Section 4.9) and Finance and Financial Markets (Section 4.10), followed by a conclusion (Section 4.11).

4.2 INDUSTRY-WIDE BENEFITS

The obvious major benefit of big data in each and every operation would be the management and analysis of the majestic amount of seismic data that is collected while operating through all the operations (through sensors, external sources and company archives). This is vital to the entire industry as for instance, a typical drilling process of deep-water oil well costs more than $100 million, which is an expensive work to undertake. Another instance would be for geologists as big data also helps in proper data management, forthcoming analysis and economic investments' assessment pertaining to all the associated aspects.

Thus, it is very important to make sure the analysis results from the data collected are precise, in order to reduce risks while saving money and time. For instance, Shell uses fibre-optic cables (created in partnership with Hewlett-Packard) to obtain precise seismic data which is then transferred to private servers maintained by Amazon Web Services. This provides all the aforementioned benefits to Shell with the help of big data.

The implementation of big data also brings optimization with it, as the use of tools and technologies pertaining to it provides deep and meaningful insights that reveal several details that (upon solving) can benefit in small yet collectively tremendous

DOI: 10.1201/9781003185710-4

amount of optimizations in almost every facet of petroleum activities (potentially saving lives and fatal injuries). These optimizations result in improved efficiency, time conservation and better economic viability of associated operations.

The insights gained from sensory stream of data also help predict the conditions that will be necessary or ideal for effective working of petroleum professionals currently and in future. This is extremely beneficial for everyone involved as knowing "what's good" helps in various areas which will be elaborated in upcoming specific sections.

Big data helps in risk reduction in almost every facet where it is applied as well. The fact that the insights gained from data analysis on sensory big data result in proper understanding of all aspects of petroleum operation ensures that a majority of potential hazards are reduced (if not eradicated). Such reduction of hazards, coupled with several others alike, ensures a hassle-free execution and working of all operations that eventually result in increased performance and economic viability for the associated parties.

It also helps in substantial reduction in the amount of carbon footprint that a company leaves through its operations. The humongous amount of data can provide insights on the aspects of operations that produce the most amount of footprint and aid in optimization (if not removal) of such operations to make significant positive impact on the environment. This notion has been popular for few years and has been adopted by several companies. For instance, Shell's 2018 sustainability report states that the company "supports the vision of a transition towards a net-zero emissions energy system" by reducing its net carbon footprint by 50% by 2050 [1] using storage technology empowered by big data software [2].

Another major upside of big data resides in its ability to extend the lifespan of companies' expensive equipment by employing various aspects of huge sensory datasets into use with the help of big data analysis. This analysis helps companies to understand and pinpoint on the aspects of operation that are working inefficiently and are under-performing. Once this is figured out, the company can proactively determine scheduled maintenance of necessary equipment to boost their performance as well as lifespan while ensuring that the least amount of maintenance-related hurdles or stops occur during operations. As an example, Shell was able to save $1 million by simply leveraging IoT data analysis in Nigeria alone [3].

One of the core benefits of using big data in any field is the predictive capabilities that it brings along with it due to technologies such as Artificial Intelligence and associated predictive models. This is extremely helpful in almost each aspect of petroleum operations as its understanding can be made as precise as possible, by feeding massive amount of seismic big data to the AI model(s) that then provide an accurate picture of "what's up" pertaining to each operation, which then provides additional benefits discussed in previous paragraph(s).

Moreover, due to its prediction capabilities, big data also provides direct benefits such as prediction of operational performance (through big data-based AI-employed analysis) and improvement of production output (via big data tools and technologies). These benefits directly impact industry outcomes such as timeline, environmental impact and profitability, while aiding in product-to-profit ratio improvement, high-frequency time series prediction, malfunction forecasting, optimal working condition prediction and carbon footprint prediction.

On a side note, an interesting yet intuitive benefit of big data implementation is the real-time and dynamic analysis that is achieved by feeding a constant stream of data into big data tools that yield constant (and real-time) analysis results, alike time series prediction in stock market automated trading. This is surprisingly handy as it helps in understanding the intricate details of petroleum processes in different scenarios (such as different periods of year, subterranean activity and oceanic and climatic conditions) and pre-determine how to handle them in future. This is a massive benefit as it helps in making real-time decisions on real-time hazardous scenarios, especially in health and safety perspective. Moreover, this is also surprisingly handy because it helps in understanding the intricate details of financial markets and make industry-wide decisions based on the same. Furthermore, all of these benefits are summarized in Figure 4.1.

4.3   BENEFITS OF BIG DATA IN GEOSCIENCE

The insights gained from sensory data also help predict the conditions that will be necessary or ideal for effective working of geoscientists. This is beneficial as knowing "what's good" helps in four major areas:

- reduction of carbon footprint of activities (for positive impact on environment)
- increase in equipment lifespan (due to various reasons such as equipment change, optimization and/or usage strategy)
- work during non-hazardous duration(s) of the day
- comprehension of expected study output (for effective and constructive decision making)

Interestingly, big data proves to be a major helping hand in making every facet of geological surveys feasible [4] including:

- improvement of data discoverability (especially survey data)
- betterment of compilation and interoperability

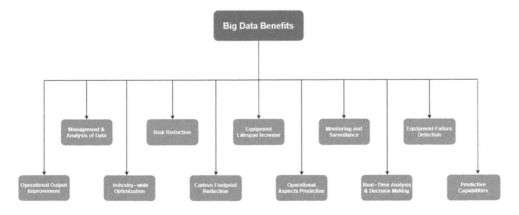

Figure 4.1 Industry-wide benefits of big data.

- geological n-D maps based on amalgamated data
- understanding of the long-term development of resources
- better mapping and modelling of geological resources

4.4   BENEFITS OF BIG DATA IN EXPLORATION AND DRILLING

A big data employed seismic analysis can also help companies to understand the nuances of a particular drilling site to make constructive and educated decision on whether to drill at that site or not based on analysis outcomes. Another aspect that seismic data analysis in exploration can help in is the understanding it can provide in terms of expected raw hydrocarbon output once operation do take place (when it is combined with other data sources like former drilling operations and research data as well).

Big data's aid in risk reduction in exploration and drilling is of massive value. In exploration, the fact that the insights gained from seismic big data can help in proper understanding of all layers beneath the surface reduces a majority of potential hazards that could have occurred during the drilling process. Moreover, the analysis of drilling data obtained from constant drilling operations also helps in keeping a check on the health of associated equipment, further reducing any hazards that might occur during the extraction process from drilling.

Big data tends to massively aid the drilling operations and associated processes as well, through various tools of which predictive modelling happens to be the major one. The integration of Artificial Intelligence and prediction models can pose major benefits in predicting various aspects such as apparent performance, equipment failure and risk management. As a typical drilling platform contains 80,000 sensors that produce 15 PB (15 million GB) in their lifespan, the analysis is generally precise and intricate with desirable analysis outcomes.

Moreover, the forecasting of equipment failure happens to be one of the most vital and deemed important aspects of big data's employment in drilling operations. The process for the same starts with deployment of sensors into the necessary location in or around drilling operation(s) where the sensory data is fed (along with equipment metadata like operation setting and model information) into AI/Machine Learning (ML)/Deep learning (DL) models, which understand and identify patterns that might cause breakdowns. This helps to predict when the equipment will fail and require complete replacement or maintenance, and thus schedule it properly to avoid any mishaps. All this generally results in reduced equipment-based downtime, prolonged equipment lifespan, hassle-free operational working and eventually save time, efforts and costs.

Furthermore, the big data collected from sensors also helps in constant monitoring and surveillance of each aspect of drilling operations (especially the equipment involved) in order to ensure safety from risks (of workers' life hazards as well as company economics) that might be detrimental to company's operations, deadlines, reputation and economic goals.

Another benefit that it provides in drilling operations is the ability to make real-time decisions. These decisions are based on the analysis insights that big data analytics provide by processing real-time data on constant basis using predictive modelling. Once the constant and live insights come up, the team associated with it (or supervisors) can make informed and effective decisions.

4.5   BENEFITS OF BIG DATA IN RESERVOIR STUDIES

The insights gained from seismic data also help predict the conditions that will be necessary for actual production of hydrocarbons. This is extremely beneficial as knowing "how everything will go" helps in four major areas:

- reduction of carbon footprint (for positive impact on environment)
- increase in equipment lifespan (due to various reasons such as equipment change, optimization and/or usage strategy; for eventually increasing the profit margins)
- have proper safety configurations in place before undertaking the activities
- comprehension of expected raw hydrocarbon output (for effective and constructive decision making on upcoming operations)

Such benefits help avoid legislative regulation, set positive public outlook, increase profitability and ensure knowledgeable process of decision making.

Alike exploration and drilling, risk reduction provided by big data in reservoir engineering helps in various facets. The insights gained from big data analytics and analysis of the reservoir result in proper understanding of intricate details of the layers beneath the earth, ensuring that a majority of future potential hazards are reduced (if not eradicated). Furthermore, big data employed seismic analysis also helps companies to comprehend the nuances of every reservoir-based operation beforehand, resulting in constructive and educated decision making along with drastic reduction in associated risks.

4.6   BENEFITS OF BIG DATA IN PRODUCTION AND
TRANSPORTATION

The improvement in the logistics of the transportation operations can be undertaken by solving complex problems such as on-site movement enhancements, vehicle route discovery, route optimization and energy consumption reduction among various others. Given the complexity of the logistics of valuable and hazardous materials like petroleum, the employment of big data can significantly help in reducing the complexity of the operations and optimizing expenditure.

Furthermore, as discussed in previous sections, the insights gained from seismic data along with real-time production sensory data can help predict the conditions that are and will be necessary for actual production of hydrocarbons (currently and in future). This is extremely beneficial as knowing "how everything is going and will go" helps in four major areas:

- reduction of carbon footprint (for positive impact on environment)
- increase in equipment and vehicle lifespan (due to various reasons such as equipment change, optimization, route efficiency improvement and/or usage strategy; for eventually increasing the profit margins)
- device optimal production strategies and transportation routes
- comprehension of expected raw hydrocarbon output (for effective and constructive decision making)

Such benefits help avoid legislative regulation, set positive public outlook, increase profitability and ensure knowledgeable process of decision making.

Big data's risk reduction in production and transportation operations with data analytics and analysis on sensory big data ensure effective understanding of the extraction process as well as transportation process in various forms (such as flow, location, temperature, abnormalities and pressure, among others), which eventually result in increased performance and profits for the companies.

4.7 BENEFITS OF BIG DATA IN PETROLEUM REFINERY

The insights gained from sensory data also help predict the conditions that will be necessary or ideal for effective working of a refinery. This is extremely beneficial as knowing "what's good" helps in four major areas:

- reduction of carbon footprint (for positive impact on environment)
- ensuring of governmental regulations in real time (to avoid legislative issues in future)
- increase in equipment lifespan (due to various reasons such as equipment change, optimization and/or usage strategy; for eventually increasing the profit margins)
- comprehension of expected refinery output (for effective and constructive decision making)

Such benefits help avoid legislative regulation, set positive public outlook, increase profitability and ensure knowledgeable process of decision making. Furthermore, big data also aims to provide further benefits for refineries that comprise of:

- improved refinery (and asset) management
- assistance in standard chemical composition of the finished products
- reduction of overall site maintenance cost
- optimization of pricing along with minimization of financial risk
- ensuring of better regulatory compliance
- determination of refined product movement in the end

4.8 BENEFITS OF BIG DATA IN NATURAL GAS

Big data aids in the improvement of logistics pertaining to all operations from extraction to decommissioning, by solving complex problems such as on-site movement enhancements, vehicle route discovery, route optimization and energy consumption reduction among various others. Given the complexity of the logistics of valuable and hazardous materials like natural gas, the employment of big data can significantly help in reducing the complexity of the operations and optimizing expenditure of execution.

The insights gained from sensory data also help predict the conditions that will be necessary or ideal for effective working of all operations (especially extraction and transportation). This is extremely beneficial as knowing "what's optimal" helps in four major areas:

- reduction of carbon footprint (for positive impact on environment)
- increase in equipment lifespan (due to various reasons such as equipment change, optimization and/or usage strategy; for eventually increasing the profit margins)
- device optimal transportation routes and methodology (that saves massive expenditure costs)
- comprehension of expected natural gas output (for effective and constructive decision making)

Such benefits help avoid legislative regulation, set positive public outlook, increase profitability and ensure knowledgeable process of decision making.

4.9   BENEFITS OF BIG DATA IN HEALTH AND SAFETY

The insights gained from sensory data also help predict the conditions that will be necessary or ideal for effective working of any operation. This is extremely beneficial as knowing "what's good" in terms of health and safety helps in four major areas:

- reduction of carbon footprint (for positive impact on environment)
- improved safety working conditions
- reduced risks and potential hazards
- decrease average worker mishaps

Such benefits help prevent health hazards, reduce unpredictability, avoid legislative regulation, set positive public outlook, increase profitability and ensure knowledgeable process of decision making. Furthermore, there are other subtle benefits [5] of big data's implementation into health and safety that together result in massive upsides:

- reduction of medical errors
- preventative methodologies along with accurate treatment
- treatment cost prediction along with treatment risk prediction
- spread of diseases modelling
- diseases detection at their early stage
- real-time alert systems
- improved healthcare staff management
- cost reduction with better efficiency

4.10   BENEFITS OF BIG DATA IN FINANCE AND FINANCIAL MARKETS

Big data aids tremendously in error reduction in petroleum finance and financial markets. Considering the fact that the insights gained from data analysis on time series big data result in proper understanding of all aspects of finance and financial markets ensures that a majority of potential analysis or calculational errors are reduced if not completely eliminated (with ability to handle exceptions). Such reduction of errors and exception handling ensures a hassle-free execution and working of all operations' financial aspect, which eventually result in increased performance (of personnel involved) and profits for the companies.

Another benefit of big data's implementation in petroleum finance and financial markets is the ability of it to provide capability. This translates to the fact that it provides the capability to perform:

- high-frequency trading
- deep learning model deployment (as DL models usually fail in the absence of huge datasets which is not an issue with big data)
- different source (platforms, services, etc.) and type (numeric, textual, etc.) amalgamation-based predictive modelling

4.11 CONCLUSION

Following the benefits discussed in this chapter, big data–employed analysis and analytics surely allow for a deeper look into data that is otherwise meaningless to provide upsides to its implementation that:

- reduce costs
- decrease time
- improve efficiency
- decrease environmental impact
- ensure upholding of regulations

among various others that have led to its widespread application in various fields of petroleum (discussed in the next chapter). Exploiting the best of the capabilities of big data, O&G companies can see drastic improvements which would else seem impossible to achieve.

REFERENCES

1. Shell (2018). Net Carbon Footprint. Shell Sustainability Report 2018. https://reports.shell.com/sustainability-report/2018/sustainable-energy-future/net-carbon-footprint.html.
2. Bekker, A. (2020). How to Benefit from Big Data Analytics in the Oil and Gas Industry? https://www.scnsoft.com/blog/big-data-analytics-oil-gas.
3. Shell Saves $1 million Using IoT (2016). Internet of Business: IoT, AI, Data and Edge Computing in the Connected World. https://internetofbusiness.com/shell-reportedly-saves-1m-using-iot/.
4. Stephenson, M. H. (2019). The uses and benefits of big data for geological surveys. *Acta Geologica Sinica*, 93(1), 64–65. http://doi.org/10.1111/1755-6724.14247.
5. Vyslotskyi, A. (2020). How to Make the Best of Big Data in Healthcare: Benefits, Challenges, and Use Cases. *N-iX*. https://www.n-ix.com/big-data-healthcare-key-benefits-uses-cases/.

Chapter 5

Applications of Big Data

5.1 INTRODUCTION

Big data refers to a set of processes such as data acquisition, data storage, search, analysis and visualization, whose cumulation can result into deep understanding of data at hand in order to make meaningful and educated decisions in the real world. While these results pose massive benefits, it is also vital to understand different applications of the same that are the reason for the boom in big data implementation.

Major applications of big data implementation include modelling, process optimization, environmental impact reduction and abiding of governmental regulations. The deeper dive into all the applications will begin with the next section. This chapter outlines the industry-wide applications (Section 5.2), applications of big data in Geoscience (Section 5.3), Exploration and Drilling (Section 5.4), Reservoir Studies (Section 5.5), Production and Transportation (Section 5.6), Petroleum Refinery (Section 5.7), Natural Gas (Section 5.8), Health and Safety (Section 5.9) and Finance and Financial Markets (Section 5.10) followed by a conclusion (Section 5.11).

5.2 INDUSTRY-WIDE APPLICATIONS

The major application of big data happens to be process modelling (including reservoir, refining, treatment, financial and comprehensive operational modelling) that helps in understanding the geological area and reservoir (and its characteristics) into consideration in a visual manner through 1D, 2D or 3D (high and multi-dimensional) geological maps based on visualization of the datasets to make better and knowledgeable decisions. The maps provide visual representation of various geological area characteristics like geological patterns, thermal signatures, depth level and depth differences that benefit petroleum engineers, data scientists and even executives (who do not necessarily understand big data and analysis insights) in comprehending the geological site and intricate working of operations in an effective manner [1].

Big data-enabled process modelling of each aforementioned type also works in other applications including reduction of environmental impact that most petroleum operations cause (as big data removes the need for hit-and-trial operations that have been under scrutiny for negative environment impacts), mitigation & ensuring of non-surpassing of regulatory limits set by governing entities (through constant stream of sensory data from the sensors in real-time) and detection of invisible non-production

DOI: 10.1201/9781003185710-5

time (through big data mining even before actual operations begin that can massively push the speed of operation).

Such applications provide optimization (by efficiency improvement), risk reduction, performance improvement(s), avoidance of massive legal battles, wastage of precious resources (that could be spent elsewhere, say quality equipment) and great deal of time [2–4]. Moreover, the said optimization (by efficiency improvement), risk reduction and performance improvement(s) that big data provides in the operations are extremely valuable to O&G companies beyond economic gains.

It also helps in the effective implementation of predictive maintenance software that analyses the aforementioned sensory big data from operation-based gear sensors to detect abnormalities (such as fatigue cracks, fault detection, safety hazards identification and stress corrosion) before they reach a critical level. This collectively helps in improved supply chain management and reduced energy consumption, through a close outlook and preventive maintenance of the underlying equipment. Therefore, it helps in avoidance of situations that can drastically impact the viability of operations beforehand [5], through a close outlook and preventive maintenance of the underlying equipment which then provides enormous financial upsides.

The employment of big data also helps in consistent and constant maintenance and monitoring of all the ongoing activities in real-time (including environmental monitoring, infrastructure maintenance and working environment [6]) while ensuring that everything works seamlessly. It is also obvious to note that this maintenance and monitoring also helps in prevention of hazards with leak detection, automated alarms, fault/breakage detection and malfunction alarms among others [6,7].

A straightforward application of big data happens to be in the analysis of seismic data that is collected (from historical databases) and generated (from deployed sensors) for effective understanding of the geology involved during the operations, especially the exploration phase. Another application for the same happens to be "Intelligent Oilfield 2.0"; for effective understanding of all activities happening in the petroleum operations from upstream to downstream. A successful implementation of big data on seismic analysis and intelligent oilfield can help companies better understand every aspect of "what's going on" and help involved personnel to make constructive financial decisions based on insights they gain from the analysis outcomes (without the need for hit and trial), thus saving time, money (massively) and efforts in the process of doing so [8–10].

Moreover, the actual process of petroleum operations can also be understood and optimized via big data, through big data–enabled simulation of fluid behaviour, by mimicking the complex mathematics- and physics-based dynamics involved in petroleum operations. This greatly helps in understanding the facility performance under various operating strategies while reducing all forms of associated safety hazards.

As an extensively mentioned attribute, big data's inherent optimization in all the operations where it is employed is a noteworthy application as well. Due to the collection of massive amounts of geologic, operational & performance data, the analysis performed on the same helps in making several major and minor revamps in each aspect of petroleum professionals' activities (like exploration improvement, refinery modelling, transportation routes, drilling, equipment update, maintenance or discard, financial modelling and time series prediction) to ensure that the efficiency of each aspect is maximized while reducing risks, downtime and heavy failure-aftermath expenditure and the amount of errors along with it, resulting in optimization of everything involved in big data's implementation.

This is especially helpful as constant monitoring, optimization and maintenance of equipment can bring forth tremendous benefits in terms of performance and cost-effectiveness. These benefits can simply be felt by understanding the fact that even a drop of 0.5% in efficiency (say) in a natural gas compression operation can cost $180,000 per annum to an O&G company in lost revenue [11].

Moving forward, big data can also be applied in the designing process of O&G fluids that can be simulated as mentioned previously. This drastically helps in improving the traditional flow of operations' comprehension, which is generally based on trial and error, while also being difficult and tedious. This is solved using artificial intelligence on big data that improves the effectiveness of the process while reducing time, difficulties and potential safety risks involved in it.

The big data-enabled simulation of fluid behaviour of the actual process of petroleum production for proper comprehension as well optimization of the operations, by mimicking the complex mathematical and physics dynamics involved in petroleum reservoir system (including reservoir rock & fluids, aquifer, surface and subsurface facilities) [12,13] is also a vital one. This greatly reduces safety hazards along with aforementioned performance-based benefits.

It includes simulation of the design of fluids such as drilling fluid, spacer, cement slurry and fracturing fluid as well, which is generally based on trial and error while also being difficult and tedious. This is solved using artificial intelligence on big data that improves the effectiveness of the process. This greatly helps in understanding the current and future production performance under various operating strategies while decreasing all forms of associated safety hazards, reducing time and the difficulties involved in it.

Further down the road, when proper understanding of the reservoir is done, big data also helps in development, optimization and monitoring of operations that aim to extract hydrocarbons beyond the norm such as enhanced oil recovery, which is a tertiary recovery method (like thermal, gas and chemical injection into reservoirs). Moreover, as 70% of the world's O&G supply comes from mature wells [14], and due to the fact that big data has shown potential to effectively implement techniques like enhanced oil recovery, it helps in increasing the productivity of mature wells (even the ones that previously seem to have reached their peak) [15] and maximizing the recovery of hydrocarbons for economic viability.

On an ending note, big data's push into any industry or field, and certainly each operation, helps massively in the development push for new scientific models using high-performance computing employed on datasets made up of historical data & real-time collected sensory, operational and financial data (like Mud Logging, Measurement While Drilling, Logging While Drilling and Gamma Ray Logging [16,17]), which again poses the benefits (including optimization, performance improvement and increased profits as mentioned in the previous chapter) when implemented properly in a precise manner as well as scenario. Moreover, an elaborated list of these applications comprising of aforementioned applications is provided in Figure 5.1.

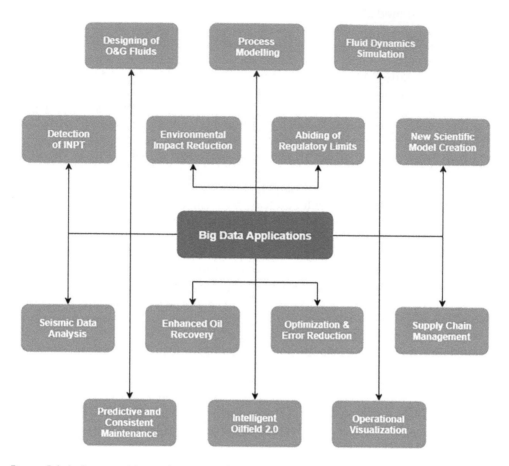

Figure 5.1 Industry-wide applications of big data.

5.3 APPLICATIONS OF BIG DATA IN GEOSCIENCE

Once the comprehension of the geological location is completed successfully as per industry-wide applications, the companies can use big data to effectively estimate various statistics around the associated processes' output results through the employment of tools and technologies at its disposal; and make knowledgeable decisions on various aspects of operation(s). Although estimations existed before the advent of big data, the integration of it into the estimation process definitely increases the accuracy of the estimated amount in less period of time.

Given the mathematical and statistical capabilities of big data, it helps in determination of several aspects of operations in a "hassle-free", "time-efficient" and "accurate" manner for activities pertaining to geoscience, including

*   Evaluation of a timeframe to feasible outcome
*   Designing of related studies
*   Development of surveillance system and raw dataset analysis

5.4   APPLICATIONS OF BIG DATA IN EXPLORATION AND DRILLING

In exploration and drilling's scenario, once the geology has been comprehended successfully, the companies can effectively locate the source of hydrocarbons and make knowledgeable decisions on where to fix the drilling site and end the exploration operation(s). On a side note, the analysis also helps in characterizing the basins that might turn out to be helpful in future for a variety of reasons including site reconsideration and data sale among other proprietary reasons.

Big data also aids in efficiency improvement of the drilling operations through drilling data that helps tremendously in terms of saving time, improving speed and maximizing profits, while making sure that least number of hazards take place (ensuring safety and risk-free environment). It further helps in prolonging the lifespan of the drilling rigs as well, which also turns out to be economically viable for the companies [18].

It helps in boosting performance of rigs by optimizing several aspects of them with the help of insights gained by drilling's big data analysis as well. The constant updates made by the analysis can constantly keep a check on maintaining optimum working environment for drilling rigs (and again, reducing costs, time and hazardous risks) while reducing the need for maintenance or equipment discard(s) [19].

One typically unknown application also happens to be the betterment of new well delivery when setting up new wells, using micro-seismic 3D imaging created via big data analytics over fibre-optic cables [20,21]. The analytic insights from big data can tremendously boost the effectiveness of managing all the processes involved in drilling and connecting a new well. The improvements brought by big data help in reducing the time lags and parallel-running well processes by a significant amount (which would otherwise not be feasible).

Given the mathematical and statistical capabilities of big data, it helps in determination of several aspects of operations in a "hassle-free", "time-efficient" and "accurate" manner for activities pertaining to exploration and drilling, including

- Evaluation of a timeframe to exploration to discovery
- Development of surveillance system
- Raw dataset analysis along with n-D mapping

Moreover, despite the fact that safety and risk reduction have been mentioned dozens of times before, it is necessary to include it as a solid application of big data in exploration and drilling operations. Considering that 60% to 80% of total industrial accidents are caused due to human error, the employment of big data–enabled systems (along with pre-determined anomaly detection) can help in severely reducing such accidents, increasing workers' safety and creating a risk-free working environment [22].

5.5   APPLICATIONS OF BIG DATA IN RESERVOIR STUDIES

Considering the fact that well testing comprises of working with data for qualitative and quantitative characterization of hydrocarbon reservoirs, the introduction of big data tools and analytics can significantly optimize and improve the process in order to gain accurate insights into identification of well test interpretation models and estimation of reservoir parameters, which happen to be its two important parts [23].

Given the mathematical and statistical capabilities of big data, it helps in determination of several aspects of operations in a "hassle-free", "time-efficient" and "accurate" manner for activities pertaining to reservoir studies, including

- Evaluation of a timeframe to production (economic life)
- Designing of petrophysics studies
- Characterization of reservoir (for current or future consideration)
- Development of surveillance system (if operations do take place)

On a side note, a reservoir modelling or simulation consists of three main parts: a geological model (which describes given porous rock formation), a flow model (which describes the flow of fluids through the porous medium) and a well model (which describes the flow in and out of the reservoir) [17]. The model's workflow goes from defining of objective to final reporting (portrayed in detail in Figure 5.2 [13,24]), and helps in optimization of development plans for new fields as well as provides assistance with operational & investment decision-making process(es).

Moving onto another reservoir modelling–based application, big data also applies in the designing process of petroleum fluids that can be simulated as mentioned previously. The traditional flow based on trial and error is solved using artificial intelligence on big data (under reservoir engineering), which improves the effectiveness of the process while reducing time and the difficulties involved in it.

Moreover, once the geology has been comprehended successfully, the companies can effectively locate the source of hydrocarbons and make knowledgeable decisions on where to fix the drilling site and end the exploration operation(s). Although pertaining to exploration and drilling, such applications come under reservoir study as

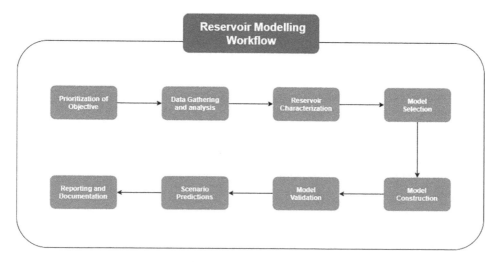

Figure 5.2 Reservoir modelling workflow.

everything related to understanding of a reservoir (including hydrocarbon discovery) falls under it. On another note, the said analysis also helps in the characterization of reservoirs that might turn out to be helpful in future for a variety of reasons including site re-consideration and data sale (among other proprietary reasons).

5.6   APPLICATIONS OF BIG DATA IN PRODUCTION AND TRANSPORTATION

Big data analytics on sensory big data produced from transportation operations helps in reducing the complexity of logistical execution of the operation by a drastic factor. Moreover, making sure that this happens while reducing the risk associated with it as much as possible is an added application/solution to an already difficult problem. Ensuring the safety of petroleum, equipment and workers involved is a major concern of O&G companies, and thus this happens to be a major application of big data.

Big data analytics also helps in reducing the energy consumption of the equipment and transportation vehicles that eventually save massive amount of expenditure on energy supply. It also helps in reducing the amount and frequency of downtimes, shutdowns or outages (as much as 36% decrease) by pre-determination of the same (through predictive and data-driven approach [25]) causing further reduction in expenditure on maintenance costs, increasing the safety of operations and improvement of asset management [5].

Given the mathematical and statistical capabilities of big data, it helps in determination of several aspects of operations in a "hassle-free", "time-efficient" and "accurate" manner for activities pertaining to production and transportation, including

• Evaluation of a timeframe for production (economic life)
• Defining of efficient transportation routes
• Error detection
• Development of surveillance system

Another extensively mentioned application of big data is the predictive capabilities that it provides to each activity performed in the operations, including predictive hydrocarbon recovery performance, predictive fault/error detection, predictive management & maintenance, predictive route optimization and Machine Learning-based predictive optimization results. This is possible due to the implementation of big data tools (artificial intelligence and machine learning) on the humongous data produced by production and transportation gear, resulting in better optimization, increased efficiency, reduced time lags and improved economic viability.

Furthermore, big data helps significantly in constant real-time understanding of the production well, resulting in aid for enhanced recovery of hydrocarbons as suggested by the sensory big data–based data analytics. This includes the understanding of storage layer and production area with high productivity, which is of high importance for O&G companies. Big data's involvement in integrating and analysing the seismic, drilling and production data for improved recovery ratio is a major economic application.

5.7 APPLICATIONS OF BIG DATA IN PETROLEUM REFINERY

Given the mathematical and statistical capabilities of big data, it helps in determination of several aspects of operations in a "hassle-free", "time-efficient" and "accurate" manner for activities pertaining to petroleum refinery, including

- Evaluation of a timeframe to finished product
- Designing of petrophysics studies
- Development of surveillance system
- Raw hydrocarbon constituent analysis

Furthermore, it even applies to concentrated activities [26] such as:

- Petrochemical asset management
- Process workflow investigation
- Distillation, thermal cracking and catalytic cracking process among others
- Energy efficiency analysis
- Prognostic analytics
- Data envelopment analysis and principal component analysis

5.8 APPLICATIONS OF BIG DATA IN NATURAL GAS

Once the comprehension of the associated operations is completed successfully, the companies can use big data to effectively estimate various statistics around the operation processes' natural gas output and make knowledgeable decisions on various aspects of operation(s). This is one of the major functions of big data–employed petroleum natural gas operations.

Given the mathematical and statistical capabilities of big data, it helps in determination of several aspects of operations in a "hassle-free", "time-efficient" and "accurate" manner for activities pertaining to natural gas, including

- Evaluation of a timeframe to finished product
- Designing of petrophysics studies
- Error detection
- Development of surveillance system
- Gas constituent analysis

5.9 APPLICATIONS OF BIG DATA IN HEALTH AND SAFETY

The actual process of petroleum operations such as refining, reservoir study and exploration can also be understood and optimized via big data, through big data–enabled simulation of fluid behaviour, by mimicking the complex mathematical and physics dynamics involved in the operations. This greatly helps in understanding the operational performance under various operating strategies while reducing all forms of associated safety hazards.

Given the mathematical and statistical capabilities of big data, it helps in determination of several aspects of operations in a "hassle-free", "time-efficient" and "accurate" manner for activities pertaining to health and safety, including

- Surveillance system development
- Maintenance process(es) setup
- Environment impact-check system development & maintenance

5.10   APPLICATIONS OF BIG DATA IN FINANCE AND FINANCIAL MARKETS

Big data helps meticulously in the asset management operations in entire industry through complex yet (almost always) efficient asset allocation and control in each operation through massive calculative as well as predictive capabilities of big data–based analysis. While a manual analysis or traditional software-aided asset management might take unfeasible amount of time to execute and still possess chances on in-effectiveness (not the best management), big data–enabled asset management can undertake the task within a fraction of previous time with accuracy unheard of.

A major application of big data is ever-so-increasing accurate financial market prediction based on humongous time-series data publicly available to everyone with an internet connection. The big data–employed machine learning (or even deep learning) models have evolved dramatically throughout the last decade and now provide astonishingly accurate forecasting capabilities, which can be economically beneficial to O&G companies on several fronts.

Big data also helps in assessing the viability of almost every petroleum operation or sub-process in economic terms. Due to its mathematical capabilities mentioned before, big data can meticulously understand a complete operation or process (from its raw materials to timeframe to finished product selling power) to understand whether that endeavour is worth taking into consideration or not (purely based on economic terms).

Given the mathematical and statistical capabilities of big data, it helps in determination of several aspects of operations in a "hassle-free", "time-efficient" and "accurate" manner for activities pertaining to finance and financial markets, including

- Hidden time series pattern detection
- Data stream amalgamation (for dynamic feeding to Artificial Intelligence model)
- High-frequency transaction feasibility

Another similar application is the use of big data by companies to effectively estimate various statistics around finance and financial markets' processes through employment of tools and technologies at its disposal; and make knowledgeable decisions on various aspects of operation(s). This happens to be one of the major functions of petroleum–related finance. Although estimations existed before the advent of big data, the integration of it into the estimation process definitely increases the accuracy of the estimated amount in less period of time.

Big data can also turn out to be useful in managing the flow of processes, and in terms of petroleum finance and financial markets, the distribution management of finished products in order to gain maximum return from the same materials. This can be made possible by analysing market trend for the said products and then sell/not-sell products based on the analysis outcomes. Besides, big data also helps significantly in marketing plan and assessment along with aid in actual product distribution through various channels in an effective, timely and non-costly manner.

5.11 CONCLUSION

Following the applications discussed in this chapter that provide the benefits provided in the previous chapter, big data–employed analysis and analytics have a wide array of applicability on varied fields that provides upsides such as reduce costs, decrease time, improve efficiency, decrease environmental impact and ensure upholding of regulations among various others. Exploiting the best of the capabilities of big data, O&G companies can see drastic improvements which would otherwise seem impossible to achieve.

REFERENCES

1. Olneva, T., Kuzmin, D., Rasskazova, S., & Timirgalin, A. (2018). Big Data Approach for Geological Study of the Big Region West Siberia. *SPE Annual Technical Conference and Exhibition 2018*, 1–6. https://doi.org/10.2118/191726-ms.
2. Yin, Q., Yang, J., Zhou, B., Jiang, M., Chen, X., Fu, C., Yan, L., Li, L., Li, Y., & Liu, Z. (2018). Improve the Drilling Operations Efficiency by the Big Data Mining of Real-Time Logging. *SPE/IADC Middle East Drilling Technology Conference and Exhibition 2018*, 1–12. https://doi.org/10.2118/189330-ms.
3. Farris, A. (2012). How Big Data Is Changing the Oil & Gas Industry. Oil and Gas Industry. https://doi.org/10.1287/LYTX.2012.06.02.
4. Brancaccio, E. Big Data in Oil and Gas Industry. Oil and Gas Portal. http://www.oil-gasportal.com/wp-content/uploads/2020/11/Big-Data-in-Oil-and-Gas-Industry-si-portale.pdf.
5. Gopinath, S., & Hampton, L. (2017). Big Oil Turns to Big Data to Save Big Money on Drilling. https://www.reuters.com/article/us-usa-oil-bigdata/big-oil-turns-to-big-data-to-save-big-money-on-drilling-idUSKBN19E0DJ.
6. Dixit, G. (2017). Internet of Things in the Oil and Gas Industry. *Journal of Petroleum Technology.* https://jpt.spe.org/internet-things-oil-and-gas-industry.
7. Bakker, A. (2020). How to Benefit from Big Data Analytics in the Oil and Gas Industry? *ScienceSoft.* https://www.scnsoft.com/blog/big-data-analytics-oil-gas.
8. Lu, H., Guo, L., Azimi, M., & Huang, K. (2019). Oil and gas 4.0 era: A systematic review and outlook. *Computers in Industry*, 111, 68–90. https://doi.org/10.1016/j.compind.2019.06.007.
9. Alfaleh, A., Wang, Y., Yan, B., Killough, J., Song, H., & Wei, C. (2015). Topological Data Analysis to Solve Big Data Problem in Reservoir Engineering: Application to Inverted 4D Seismic Data. *SPE Annual Technical Conference and Exhibition 2015*, 1–17. https://doi.org/10.2118/174985-ms.
10. Roden, R., & Ferguson, J. (2016). Seismic Interpretation in the Age of Big Data. SEG *Technical Program Expanded Abstracts 2016*, 1–5. https://doi.org/10.1190/segam2016-13612308.1.
11. Gillette, B. (2017). Big Data Analytics Protects O&G Midstream Compressors. *Logilube.* https://www.logilube.com/single-post/2017/02/09/Big-Data-Analytics-protects-OG-Midstream-Compres.

12. Hutchinson, M., Thornton, B., Theys, P., & Bolt, H. (2018). Optimizing drilling by simulation and automation with big data. *Society of Petroleum Engineers*, 1–16. https://doi.org/10.2118/191427-ms.
13. Reservoir Simulation (2021). Oil and Gas Portal. http://www.oil-gasportal.com/reservoir-simulation/fundamental.
14. Hassani, H., & Silva, E. S. (2018). Big Data: a big opportunity for the petroleum and petrochemical industry. *OPEC Energy Review*, 42(1), 74–89. https://doi.org/10.1111/opec.12118.
15. Cowles, D. (2015). Oil, Gas, and Data: High-Performance Data Tools in the Production of Industrial Power. https://www.oreilly.com/content/oil-gas-data/.
16. Nicholson, R. (2012). Big Data in the Oil & Gas Industry. *IDC Energy Insights*, September 2012.
17. Seshadri, M. (2013). *Big Data Science Challenging the Oil Industry*. CTO Global Services, EMC Corporation, March 2013, Hopkinton, MA. http://web.idg.no/app/web/online/event/energyworld/2013/emc.pdf.
18. Duffy, W., Rigg, J., & Maidla, E. (2017). Efficiency Improvement in the Bakken Realized Through Drilling Data Processing Automation and the Recognition and Standardization of Best Safe Practices. *SPE/IADC Drilling Conference and Exhibition 2017*, 1–10. https://doi.org/10.2118/184724-ms.
19. Maidla, E., Maidla, W., Rigg, J., Crumrine, M., & Wolf-Zoellner, P. (2018). Drilling Analysis Using Big Data has been Misused and Abused. *SPE/IADC Drilling Conference and Exhibition 2018*, 1–25. https://doi.org/10.2118/189583-ms.
20. Bertocco, R., & Padmanabhan, V. (2014). Big Data Analytics in Oil and Gas. https://www.ogj.com/home/article/17293403/big-data-analytics-in-oil-and-gas.
21. Crooks, E. (2018). Drillers Turn to Big Data in the Hunt for More, Cheaper Oil. https://www.ft.com/content/19234982-0cbb-11e8-8eb7-42f857ea9f09.
22. Johnston, J., & Guichard, A. (2015). New Findings in Drilling and Wells Using Big Data Analytics. *Offshore Technology Conference 2015*, 1–8. https://doi.org/10.4043/26021-ms.
23. Deng, Y., Chen, Q., & Wang, J. (2000). The Artificial Neural Network Method of Well-Test Interpretation Model Identification and Parameter Estimation. *International Oil and Gas Conference and Exhibition 2000*, 1–7. https://doi.org/10.2118/64652-ms.
24. Jamshidnezhad, M. (2015). Reservoir modeling. *Experimental Design in Petroleum Reservoir Studies*, 9–56. https://doi.org/10.1016/b978-0-12-803070-7.00002-8.
25. Yuan, Z., Qin, W., & Zhao, J. (2017). Smart manufacturing for the oil refining and petrochemical industry. *Engineering*, 3(2), 179–182. https://doi.org/10.1016/j.eng.2017.02.012.
26. Patel, H., Prajapati, D., Mahida, D., & Shah, M. (2020). Transforming petroleum downstream sector through big data: A holistic review. *Journal of Petroleum Exploration and Production Technology*, 10(6), 2601–2611. https://doi.org/10.1007/s13202-020-00889-2.

Chapter 6

Implementation of Big Data

6.1 INTRODUCTION

As inferred from the massive benefits and wide array of applications of big data, it certainly helps in deep understanding of data at hand in order to make meaningful and educated decisions in the real world. This simple aspect has led to widespread implementation of big data in every industry, including oil and gas. The upcoming section aims to provide an explanation to standardized workflow of big data's implementation for the aforementioned reasons. Major workflow aspects include Key Performance Indicators (KPIs) setting, deployment, extraction, transformation and loading (ETL) and Artificial Intelligence (AI) analysis. The deeper dive into all of them will begin with the next section (Section 6.2) followed by specific implementation aspects (Section 6.3) and a conclusion (Section 6.4).

6.2 IMPLEMENTATION PROCESS

The success of big data's integration into petroleum operations is majorly dependent on its effective implementation, which consist of various steps which are hereby mentioned. The initial aspect of implementation that needs to be tackled is the defining process of the business problem to solve as well as KPIs for the same. This includes asking of the right questions for the operations(s) at hand while ensuring that the asked questions are rightfully interpreted by everyone on the team (involved petroleum engineers and professionals along with data scientists). It also portrays the setting of goals, i.e., what to expect as an outcome for employing big data along with key milestones in the complete process.

Once the business goal is set and expectations are taken care of, a team of researchers, data scientists and petroleum engineers is formulated that would undertake the project of integrating big data into various operations and associated sub-tasks. This team is supposed to be able to meet all the demands of the project (such as knowledge and experience of handling all the necessary big data tools and technologies discussed in the challenges chapter).

Moving forward, equipment and sensors' deployment phase begins in various locations at viable geological spots that would play a key role in big data analysis by generating the humongous amount of data on which the analysis (and/or analytics) is performed based on operational needs. The deployment of sensors in correct and precise locations is important as well, in order to gain quality and highly accessible (in terms of transmission

DOI: 10.1201/9781003185710-6

and availability) data for better insights. Moreover, in cases such as financial markets, this step comprises of data acquisition or setup of data pipeline that feeds a continuous stream of financial time series data for effective stock market and product price forecasting.

Once the data sources (sensors and/or equipment) and subsequent data collection tasks are taken care of, the data then undergoes rigorous processing that includes cleaning and segregation (based on data type and amount). This process of ETL usually amounts for 80% of the total time required for gaining insights with big data analytics.

This ETL process is then followed up by the actual analysis of cleaned and structured data by feeding it into AI/Machine Learning (ML)/Deep Learning (DL) models (generally on cloud platforms) that find patterns in the massive pile of data to give meaningful insights and make accurate predictions. The choosing of appropriate models for the analysis process is also vital and is either carried out by taking models' historical performance or short-scale hit-and-trial performance (in case of newer model or data) into consideration.

At the end, after the insights are gained and are deemed to be of satisfactory level, the results are portrayed to company executives with an interactive, clean, simple and feasible user interface. This is important as the feasibility of result's showcase on the interface aids the executives in petroleum engineering-based decision-making process effectively. Moreover, Figure 6.1 portrays a summarized visual workflow of the entire big data implementation.

Furthermore, a successful implementation of big data into petroleum operations, as per a study by Cameron [1] on petroleum industry, can be measured by the following nine factors:

- Accurate definition for addressing a limited and recognizable business problem through big data
- Combination of big data methods with statistics and mathematical-based data analysis

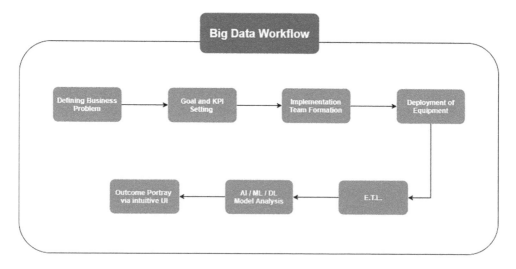

Figure 6.1 Big data workflow in petroleum operations.

- Implementation of big data using interdisciplinary team of data scientists and petroleum engineers
- Being need-driven strategy (rather than technology-driven)
- Presentation of the results using a user-friendly and aesthetically pleasing user interface
- Being qualified through a small (yet realistic) "proof of concept"
- Minimal portrayal of raw big data, and instead, portray pictures, KPIs, models and recommendations
- Address of how the solved problem is associated with the big picture (such as better financial profitability)
- Appliance-type delivery of outcomes (with clearly defined IT architecture and data store interfaces)

If the above factors are taken into consideration and upheld to their utmost potential, the implementation of big data will be considered as successful in petroleum operation(s).

6.3   OPERATION-SPECIFIC ASPECTS OF BIG DATA

Depending on the specifications (and requirements) of the operation along with its usability, there also might be a need for additional software developers/engineers who aim to aid the data scientists and petroleum engineers in turning their vision into code (or no-code platform in case of long-term). This is necessary as despite of big data platforms and data scientist's knowledge, there is always a need for SDE/SWE who connects both the worlds for effective interconnective working of them.

Another implementation aspect of big data that is dependent on the specifications (and requirements) of the operation is the creation of an operation-specific big-data architecture. This might not be necessary in case of operations where working with data might change from location to location such as exploration and reservoir studies; however, in operations like production and refining, this is certainly useful as it streamlines the process of implementation in the long term (by formalizing the data sources, both existing and potential, along with standard data flows) with massive environmental, financial and regulatory benefits.

The implementation of big data throughout the O&G company can also require the need of training sessions in case the workforce that will actively use big data solutions is big. It can take the form of workshops with Q&A sessions or instructor-led training, any form that works for the company. It essentially helps the end-users to understand (and appreciate) how to use the big data–enabled solution(s) to get valuable and actionable insights that benefit the organization [2].

An interesting thing to note about the implementation of big data is that it is not necessary to follow the workflow as it is, in certain scenarios or operations. For instance, a company might need to follow each aspect of the workflow in case implementation is based on finance and financial markets (as requirements and long-term goals are already in place). However, in operations like exploration where long-term feasibility is not defined yet (as the goal itself of the operation is to define that!), it is better off for the O&G company to implement only a certain portion of the workflow and even that, not in same flow.

6.4 CONCLUSION

Following the workflow discussed in this chapter can provide the benefits of application mentioned in the previous chapters, including better asset management, goal clarity, improve efficiency, clear set of instructions (along-with direction) and clear portrayal of outcomes among various others. These aspects streamline the operations like no other and every company (hands-down) aims to achieve the same by any means possible.

Undertaking the said implementation workflow along with operation-specific additions can help O&G companies exploit the best of the capabilities of big data that result in drastic improvements which would else seem impossible to achieve in big data's absence.

REFERENCES

1. Cameron, D. (2014). Big Data in Exploration and Production: Silicon Snake-Oil, Magic Bullet, or Useful Tool? *SPE Intelligent Energy Conference & Exhibition 2014*, 1–9. https://doi.org/10.2118/167837-ms.
2. Mikhailouskaya, I. (2019). How to Make Big Data Implementation a Success: Roadmap and Best Practices to Follow. *ScienceSoft*. https://www.scnsoft.com/blog/big-data-implementation.

Chapter 7

Big Data Platforms

7.1 INTRODUCTION

Taking the massive benefits, wide array of applications of big data and the standard implementation workflow into consideration, it is certainly useful to understand the backbone, big data platforms, that helps to make meaningful and educated decisions in the real world. The comprehension of the same helps data scientists, petroleum engineers and executives to deeply understand how the implementation is "executed" while understanding its related aspects such as expenditure, level of expertise required and execution timeframe. The upcoming section aims to provide an enumeration to standard big data platforms in use today. Major big data platforms include Hadoop, IBMPureData and Microsoft MURA. The deeper dive into all of them will begin with the next section (Section 7.2), followed by benefits (Section 7.3), roles (Section 7.5) and conclusion (Section 7.6).

7.2 PLATFORMS

As discussed in previous chapters, O&G companies have started their transition into cloud services (either built on their own or using public services such as Amazon Web Services (AWS), Google Cloud Platform (GCP) or Azure) for maintaining their humongous amount of data. However, once the data has been transferred to cloud services, it needs to be analysed as well, mostly using big data platforms. There are various platforms in the marketplace, but the following turn out to be the major and widely used ones:

- **Hadoop Infrastructure**:
    Hadoop happens to be one of the most applicable means of analysing massive amount of seismic data. It takes data as input and splits it into multiple clusters, that each can be processed individually by using several maps (reducing the number of workers).
- **IBMPureData**:
    This innovative tool by IBM lets petroleum companies to analyse huge amount of data for deep and complex big data analytics, while supporting even terabytes of data input.
- **IBM InfoSphere**:
    IBM InfoSphere helps in mitigating a vital issue with analysing humongous amounts of datasets, which is time required to process everything. Thus, while

DOI: 10.1201/9781003185710- 7

supporting massive amounts of data, InfoSphere undertakes complex computations in a comparatively less amount of time.

- **Microsoft MURA:**
    This platform developed by Microsoft aims at improving the efficiency of workers and mitigates business-related issues of petroleum companies through various aspects such as Real-Time Analytics, People-Process-Information Integration, Self-Serve Business Intelligence, Rich Interactive User Experience and Mobility.
- **Oracle Architecture Development Process (OADP):**
    OADP is an Oracle platform that's specifically created for big data analytics (especially for oil and gas companies), which consists of different products that cater to companies' needs. Few concepts in OADP architecture include Oracle Data Integrator, Oracle Real-time Decisions, Oracle Big Data Discovery, Endeca, Exadata and NoSQL Database.

These are the majorly preferred platforms (while Hadoop being most widely used). Although these platforms are small in number, the advent of big data into petroleum industry will surely give rise to other feasible platforms as well. Moreover, it is generally deemed better for companies to develop specialized big data tools that cater to their needs and that include data collection (usually sensor data or financial statements or time series data based), data storage (in on-premise facilities or cloud services) and analysis tools & technologies. This helps the companies in several ways including software ownership cost reduction while optimizing the value of collected data [1].

However, this is feasible to undertake only for companies that have massive technological investment budgets. Therefore, most petroleum companies for now will have to rely on the aforementioned platforms (until the specialization becomes feasible for everyone due to technological advances). Furthermore, the factors mentioned in the next section are the benefits pertaining to the use of platforms (like the aforementioned ones).

7.3   BENEFITS

The obvious yet major reason and benefit of using the platforms is the reduction in costs of using big data along with the ease of implementing big data into the integrated ecosystem (big data infrastructure). The platforms ensure that the O&G company's involvement is significantly less in maintaining infrastructure for big data so that it can focus more on their services. On a side note, the platforms do provide all the benefits that migration to cloud services provide on top of the industry- and/or company-specific requirements.

Another similar benefit of using the platforms is the increase in output (analyses outcomes) with less usage of resources in lesser timeframe. As using hardware-based tools require engineering input, it takes considerable amount of time. Thus, the non-requirement of infrastructure maintenance enables data scientists and petroleum engineers to focus on actual analyses process, thus giving increased output in less amount of time.

As discussed earlier, big data comes in three forms – structured, unstructured and semi-structured, and handling all such types of data can be a tedious and time-consuming task. And the entire process of segregating and managing different data types is taken care off by the platforms. They even help in integrating and combining one or more types (or even sources!) to gain high-value insights. Moreover, considering the fact that 90%

of petroleum industry generated data is unstructured [2] in nature ("variety" character-istic), the inherent capability of platforms to handle and manage different types of data can save massive amount of time and effort of the entire technical team.

Furthermore, besides the fact that the platforms provide feasibility of handling and managing big data, they also provide astonishing superior tools and technologies at the company's fingertips, making performing any kind of operation (edit, update, delete, move, segregate, analyse or visualize) easier than ever before, which can be used by individuals who have no experience into big data or engineering at all. In this manner, the big data team can perform complex and cutting-edge analyses (like deep-learning-based prediction) through the platform itself without writing a single line of code.

And finally, the platforms have capability to provide visually aesthetic portrayal of data, data analysis and analysis results without the need of code, with better precision and in less amount of time. These visualizations can then be exported to different formats (image, PDF, etc.) or directly shared on the cloud servers with the business executives. Moreover, Figure 7.1 summarizes all the platform-based benefits in a single glance.

Figure 7.1 Benefits of big data platforms.

7.4 OTHER ASPECTS

Other than the platform itself, there are various tools, methods, languages and technologies that come into play while working with them. For instance, there can be different databases (SQL, NoSQL and NewSQL), warehouses (Hive and Hadoop), data processing (batch and stream), query language (HiveA1 and Pig Latin), machine learning (R and Python) and log processing (Splunk and Loggly). Table 7.1 broadly describes all of these in a concise manner.

7.5 ROLES

Furthermore, once a platform has been set up successfully by the engineers & data scientists and is functional to use, it must accomplish the following roles for proving to be a worthwhile investment:

- It must take care of controlling real-time stream of data produced (industry sensors, financial statements and financial market), while ensuring its safety and integrity (from the Confidentiality-Integrity-Availability (CIA) triad of cybersecurity).
- Once the data has reached the platform, it must handle data engineering (or processing it in terms of cleaning, segregation and updating).
- The platform must integrate all sources of data, while performing analyses on the continuous stream of data, and provide outcomes in real-time.
- Upgrade the infrastructure hardware, technologies, tools and (especially) analysis artificial models automatically and on scheduled basis without human intervention.
- Platform should also manage and control on-premise systems in real time through real-time stream of data based on real-time continuous analysis.

If the implementation of a platform by a company takes care of these aforementioned roles, then it has fulfilled its purpose and the O&G company has made a viable and worthwhile investment that will help economically and environmentally (and event legislatively) for years to come.

Table 7.1 Category-based examples of tools, methods, technologies and languages provided by big data platforms

Category	Examples
Platform	**Local**: Hadoop, Cloudera, InfoSphere big-insights, ASTERIX
	Cloud: Amazon Web Services, Google Cloud Platform, Azure
Database	**SQL**: Greenplum, Aster data, Vertica
	NoSQL: HBase, Cassandra, MongoDB, Redis
Data warehouse	Hive, Hadoop DB, Hadapt
Data processing	**Batch**: MapReduce, Dryad
	Stream: Storm, S4, Kafka
Query language	HiveAl, Pig Latin, DryadLINQ, MRQL, SCOPE
Statistics & ML	Python, R, Mahout, Weka
Log processing	Splunk, Loggly

7.6 CONCLUSION

Following the big data platforms along with their benefits, aspects and roles discussed in this chapter, it can easily be inferred that the advances in technology and platform-based services have grown to massive extent where they cater to a huge set of problems faced by every industry, including oil and gas. Several companies in the industry have even struck deals with big data companies to provide specialized platform services for their operations.

It is now set in concrete that these platforms represent the backbone of the big data implementation in the industry, and if O&G companies exploit the best of the capabilities of them, it can result in drastic improvements (in their processes, management, regulatory avoidance, environmental impact and economic terms among several others) which would else seem impossible to achieve in big data platforms' absence.

REFERENCES

1. Preveral, A., Trihoreau, A., & Petit, N. (2014). Geographically-Distributed Databases: A Big Data Technology for Production Analysis in the Oil & Gas Industry. *SPE Intelligent Energy Conference & Exhibition 2014*, 1–10. https://doi.org/10.2118/167844-ms.
2. Ishwarappa, J., & Anuradha, J. (2015). A brief introduction on big data 5Vs characteristics and hadoop technology. *Procedia Computer Science*, 48, 319–324. https://doi.org/10.1016/j.procs.2015.04.188.

Chapter 8

AI Algorithms

8.1 INTRODUCTION

Throughout the discussion in the previous chapters comprising operations, benefits, application, implementation and platforms, a single thing that has been mentioned on a consistent basis is Artificial Intelligence (AI). AI simply refers to the simulation of human intelligence and its processes by machines, usually computer systems without any human intervention. It is essentially science (like mathematics or physics) that studies different methodologies to build intelligent programs and machines that can creatively solve problems.

Machine Learning (ML), on the other hand, is a subset of AI that enables systems to automatically learn and improve from experience without being explicitly programmed. In ML, there are different algorithms (such as neural networks) that help to solve problems. In general, ML works by feeding itself huge amounts of training data which is then analysed for correlations and patterns (among others as per the application) that help make predictions that it is known for.

And finally, deep learning is a subset of ML that uses neural networks (mimic of human brain's biological neurons) to analyse different factors with a structure that is similar to the human neural system. An interesting aspect of DL is that while standard ML algorithms are usually linear in terms of working, DL algorithms are actually stacked in a hierarchy of increasing complexity and abstraction that results in increased understanding and better predictions. The aforementioned aspects on AI can be easily understood with Euler diagram in Figure 8.1 [1,2].

The deeper dive into the major AI algorithms in the industry will begin with the next section (Section 8.2), followed by other operation-specific aspects of it (Section 8.3) and a conclusion (Section 8.4).

8.2 MAJOR AI ALGORITHMS

Although there are hundreds (if not thousands) of AI algorithms out there, which help in specific prediction and analysis scenarios (like classification, regression, decision trees, reinforcement learning, etc.), some are used more than others, and here are the major algorithms used in petroleum operations:

DOI: 10.1201/9781003185710-8

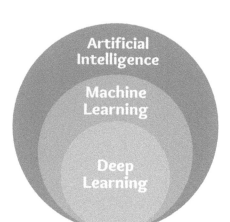

Figure 8.1 Artificial Intelligence Euler diagram.

- **Artificial Neural Network (ANN):**
    ANN algorithms have inherent ability to process complex, blended and perplexed information and are simply a series of calculations that aim to imitate the human brains' working (through mimicking cerebrum structure with artificial neurons/nodes).
- **Support Vector Machine (SVM):**
    SVM algorithm is a supervised and instance-based learning model that focuses on finding a hyperplane that creates a line of separation between different classes/types of data points for mostly classification purposes.
- **K-Nearest Neighbour (KNN):**
    KNN algorithms aim at classification of data points, while taking the parameters or value of their nearest neighbour into consideration. Its implementation generally results in decreases of probability of error occurrence [3], by accurately predicting where a data point lies in a set of similar-characteristic groups of data.
- **Decision Tree (DT):**
    DT algorithms are prediction models that use decisions based on trees to conclude a data point's value/outcome (depicted as a leaf) from its parameters (depicted by the branches). In a nutshell, it builds a model for decisions based on attributes of the training data and is used for both classification and regression purposes.

These algorithms represent the majority of Artificial Intelligence implementation upon big data in petroleum operations, while companies may use a combination of them (or even develop custom algorithms) for improved analysis in terms of accuracy and time constraints.

8.3   OPERATION-SPECIFIC ASPECTS

Few petroleum operations (such as refining, treatment, natural gas and health and safety) have seen a widespread use of Principal Component Analysis (PCA) for dimensionality reduction of humongous datasets that helps in increasing dataset's interpretability while minimizing information loss. It does so by creating new uncorrelated variables that successively maximize variance, resulting in better implementation of the aforementioned AI algorithms that in return provide improved predictions. There are other dimensionality reduction methods as well, such as:

- Missing Value Ratio
- Low Variance Filter
- High Correlation Filter
- Backward Feature Elimination
- Forward Feature Selection
- Factor Analysis
- Independent Component Analysis

However, due to the benefits that PCA imparts, it can be seen through extensive research performed in the last decades (mentioned in the research chapter) and real-world implementation (mentioned in another chapter) that PCA is the most feasible dimensionality reduction technique.

Moreover, various algorithms are used in certain operations in the oil and gas industry, which are hereby mentioned:

- **Genetic Algorithm (GA):**
  GA is an adaptive heuristic search algorithm that belongs to the larger part of evolutionary algorithms, based on ideology of natural selection and genetics. It is commonly used to generate high-quality solutions for optimization problems and search problems. Due to its capabilities, it used extensively in petroleum reservoir studies, refining, health and safety and even finance and financial markets.
- **Long Short Term Memory (LSTM):**
  LSTM Network is an advanced Recurrent ANN, a sequential network, that allows information to persist inside the network for better predictive capabilities. It is capable of handling the vanishing gradient problem faced by RNN and is generally used for time series predictions that are prevalent in petroleum finance and financial markets.
- **K Means Clustering (KMC):**
  It is basically an algorithm that works on probability where each data point is associated with a probability of belonging to another cluster. It is also known as a fuzzy algorithm as the data points do not have an absolute or concrete membership inside a particular cluster. It is usually used in classification tasks such as anomaly detection, and hence used in health and safety and finance and financial markets–related petroleum operations.

Furthermore, all the major algorithms along with the operation-specific aspects can be visualized in a single glance with Figure 8.2.

Another aspect of big data that was previously discussed but needs to be re-iterated is the fact that specifications (and requirements) of the operation do shape the way AI algorithms are implemented through creation of an operation-specific big data architecture. This might not be necessary in case of operations where the applied AI algorithms are used industry-wide. However, in specific operations like creation of an architecture

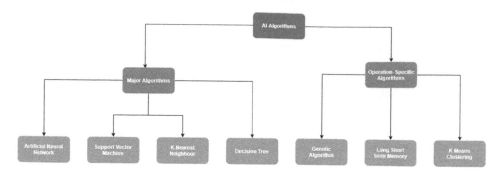

Figure 8.2 Major AI algorithms with operation-specific aspects.

for betterment of AI algorithm, analysis does surely streamline the process of implementation in the long term with huge economic upsides.

8.4 CONCLUSION

Artificial Intelligence is the backbone of the big data analysis that actually provides the benefits discussed in previous chapter, and thus, proper understanding of the AI-based algorithms is of utmost importance, which is exactly what has been discussed in this chapter. Utilizing the said algorithms (along with others in case of specific operations) can help O&G companies exploit the best of the capabilities of big data analysis that result in drastic improvements which would else seem impossible to achieve in big data's absence.

REFERENCES

1. Gavrilova, Y. (2020). Artificial Intelligence vs. Machine Learning vs. Deep Learning: Essentials. *Serokell*. https://serokell.io/blog/ai-ml-dl-difference.
2. Burns, E., Laskowski, N., & Tucci, L. (2019). What Is Artificial Intelligence? *TechTarget*. https://searchenterpriseai.techtarget.com/definition/AI-Artificial-Intelligence.
3. Murty, M. N., & Devi, V. S. (2011). Nearest Neighbour Based Classifiers. Undergraduate Topics in Computer Science, 48–85. https://doi.org/10.1007/978-0-85729-495-1_3.

Chapter 9

Research on Big Data

9.1 INTRODUCTION

A salient aspect of any field happens to be the amount of research work that is put into it, and considering the advent of big data along with the benefits it imparts on diverse applications, it is rather obvious to note that there has been a significant contribution from the research community on the research work upon big data in various associated operations.

Big data refers to a set of processes such as data acquisition, data storage, search, analysis and visualization whose cumulation can result in a deep understanding of data at hand in order to make meaningful and educated decisions in the real world. These results pose massive benefits on the application field that are the reason for the boom in big data implementation.

Major research work on big data implementation includes work on seismic analyses, anomaly detection, prediction/forecasting and optimization. The deeper dive into all the research work pertaining to industry (and subsequently each operation) will begin with the next section. This chapter outlines the industry-wide research (Section 9.2) research of big data in Geoscience (Section 9.3), Exploration and Drilling (Section 9.4), Reservoir Studies (Section 9.5), Production and Transportation (Section 9.6), Petroleum Refinery (Section 9.7), Natural Gas (Section 9.8), Health and Safety (Section 9.9) and Finance and Financial Markets (Section 9.10), followed by a conclusion (Section 9.11).

9.2 INDUSTRY-WIDE RESEARCH

Research by von Plate [1] was focused on the study of integration of all sources of data in order to utilize it for advanced analytics solutions. These solutions could then provide improved petroleum asset management and assist operators with detailed and focused insights into the future. And these insights have the potential to provide benefits such as minimized downtimes (through scoping of maintenance and appropriate long-term scheduling), proactive asset RUL (remaining useful life) management through intelligent load maintenance and optimum asset deployment (by taking future asset risk profiles into account). These benefits pertain to majority of aspects in all operations of refinery and natural gas as well, and thus the study is of vital importance [1].

Research by Anagnostopoulos [2] showed that propulsion power can help improve transportation ship performance by improving the evaluation strategy for ship(s) that

DOI: 10.1201/9781003185710-9

results in lower emissions and greener future operations. Moreover, they also conducted data analytics by using eXtreme Gradient Boosting (XGBoost) and Multilayer Perception (MLP) neural networks methods in the study, where the MLP is an updated version of scikit-learn Python Library's neural network models (big data technique). With dataset taken from over 3 months of sensory data through a large car truck carrier massive vessel, the outcomes clearly showed positive environmental impact and increased performance [2].

Big data was also used in detecting Invisible Non-Production Time (INPT) by Yin et al. using real-tile logging data, for optimizing extraction by optimizing (or removing) the said INPT through statistics, artificial intelligence (or Machine Learning/ML or Deep Learning/DL) or cloud computing [3].

A study by Azadeh et al. [4] focused on modelling of energy efficiency and petroleum refinery (including treatment for natural gas) estimation by proposing a model for energy efficiency evaluation. They used data envelopment analysis as an analysis tool for determination in energy consumption improvement & optimization, and their proposal was verified and validated by principal component analysis (PCA) and non-parametric Spearman correlation technique [4].

Another study by Wang et al. [5] carried out research on storage equipment for measuring and exploring risk-level and forewarning management using data modelling and data mining. Upon using various models such as K-means cluster analysis and logical regression fitting, they found linear regression model to be the one that provided best fitting in risk assessment of Oil and Gas (O&G) storage. Other than that, they also ideated on online & intelligent forewarning management information system which helps in management (including big data monitoring, intelligent assessment, automatically forewarning and advanced contingency) [5].

A study by Fan et al. (2019) [6] provided a comprehensive method for natural gas pipeline efficiency evaluation using energy & big data analysis. They developed a data envelop analysis model for input and output variation analysis of energy and volume, in order to determine the efficiency of natural gas pipeline. Their work was successful with conclusion on the fact that the growth of efficiency is inversely proportional to the amount of natural gas transmission while accepting that economic efficiency and transmission consist of trade-off relationship between themselves.

An interesting study by Tokarek et al. [7] used PCA analysis for measuring air pollutants from Athabasca oil sands, where they used "varimax-rotation" along with PCA to illuminate 28 variables (such as volatile organic compounds and intermediate-volatility organic compounds), while ten of them were spotted and classified based on source type for feasible usage in petroleum operations which would tremendously help in operations such as refining, natural gas and health & safety [7].

Reservoir modelling has seen tremendous amount of research work in recent years too, especially in unconventional oil and gas resources while also proving beneficial to production operations. For instance, a study by Lin [8] provided insights on principles of big data algorithms and applications for unconventional O&G resources, through a workflow of three levels, namely, computational domains & numerical schemes, fitting of pressure valves and domain & boundary conditions convergence. Another study by Chelmis et al. [9] provided a semi-automatic and semantic assistance approach for manual data curation of smart oil fields with the help of big data, which solved the problem of managing and storing humongous amounts of user-generated unstructured

Table 9.1 Research work pertaining to big data in entire industry

Research	References
Advanced analytics based petroleum asset management	[1]
Big data-based ship performance improvement study	[2]
Detection of INPT using real-tile logging data	[3]
Modelling of energy efficiency and petroleum refinery estimation	[4]
Data mining-based O&G risk assessment and online forewarning	[5]
Natural gas pipeline efficiency evaluation	[6]
PCA-based air pollutant measurement in O&G operations	[7]
Reservoir modelling for unconventional oil and gas resources	[8,9]

data with no proper metadata. Moreover, the research work mentioned until now in this section is summarized in Table 9.1.

9.3 RESEARCH OF BIG DATA IN GEOSCIENCE

A research work by Qi and Xuelong [10] was aimed at new big data methods and ideas in geological scientific research and provided a comprehensive outlook to big data in geoscience. They discussed how traditional perspectives are being challenged and subverted in the era of big data, while mentioning how big data is the step forward to traditional methods rather than complete overhaul of them. The research work also describes the current knowledge structure to be unreasonable along with the future direction of the same, which will require two kinds of experts, namely, big data and artificial intelligence professionals & geology professionals [10].

Another research work by Spina [11] was focused on capabilities and potential of big data and artificial intelligence in geoscience. The research work portrayed the fact that the addition of graphical processing unit (GPU) based processing resources to typical equipment of a server host can speed up queries performed on large databases while reducing training time for deep learning architectures. It also outlined a (then) recent real-world implementation of big data and AI into humongous dataset that led to the creation of an interactive virtual globe that showed a colour mosaic of seabed geology, to emphasize on the fact that big data-enabled analyses performed on large heterogeneous dataset(s) can help in the discovery of hidden models and unknown correlations that allow more solid reconstructions and forecasts on natural phenomena that have had and will have a major impact on our planet's ecosystem(s) [11].

Another similar study by Chen et al. [12] was dedicated on the review of applications of big data and artificial intelligence in geology. Upon rigorous analysis in the study, several key conclusions were made. For instance, it was noted that due to geological data acquisition methods' diversification, storage cost reduction and historical data accumulation, the advent of big data has become feasible for real-world implementation. Moreover, although big data and artificial intelligence were tagged to be in their initial stages in geology, several specific methods were mentioned for different aspects of geological operations (such as artificial neural network or ANNs for classification and identification of minerals). Big data was also deemed applicable to sample identification, mineralogical study and earthquake monitoring, with the ability to improve efficiency, reduce errors and improve accuracy with the help of AI [12].

An interesting study by Bruce [13] outlined the basic concepts pertaining to game theory and its mechanism's application to big data analytics for decision-making in the field of geosciences. This was followed by a proposition of use of strategic competitive game theory models for spectral band grouping when exploiting "hyperspectral imagery" using conflict data filtering and a strategy interaction process. The proposed system used Nash equilibrium for the band grouping problem and implemented the model under the assumption of all players as rational. The proposed band grouping was used as a component in a multi-classifier decision fusion system for automated ground cover classification with hyperspectral imagery. The experimental results demonstrated that the proposed game theoretic approach significantly outperformed the comparison methods, showcasing big data's supremacy [13].

In a nutshell, the research work pertaining to big data's implementation in geoscience has exploded and, due to its catered benefits, seems to keep going for at least the next few decades when big data might become the norm. Moreover, the research work mentioned until now in this section is summarized in Table 9.2.

9.4   RESEARCH OF BIG DATA IN EXPLORATION AND DRILLING

Research by Barney in 2015 [14] established the fact that exploration is one of the most significant applications of big data within the petroleum industry while discussing the role of supercomputers in seismic research!

Among all the research work into exploration, there has been a major focus on the process of finding traces or patterns of possibility of existence of hydrocarbons underneath the surface, resulting in reduced usage of materials, manpower and logistics. As an example, Feblowitz [15] utilized big data analytics (in the form of pattern recognition) on a massive seismic dataset for identifying "productive seismic trace signatures" or reservoir findings that would otherwise be overlooked. The same was also seen in a study by Soofi and Perez [16] where big data along with advanced data analytics can help identity reservoir traces. The notion was further strengthened when Baaziz and Quoniam [17] portrayed that analysis of 3D seismic data can provide an accurate outlook for the identification of hydrocarbon deposits (reservoirs).

Another research by Roden [18] focused on multi-component seismic analysis (for reservoir detection and geology understanding) using PCA along with Self-Organizing Maps (SOMs), carried out through five stages. The first stage involved defining the geological issue, followed by the second stage where key attributes related to the issue were defined (using PCA). Thereafter, SOM was run by employing machine learning tools for creating a prediction model. In the fourth stage, SOM analysis results were further analysed for gaining vital geological features, which were then refined in the

Table 9.2   Research work pertaining to big data in geoscience

Research	References
New big data methods and ideas in geological scientific research	[10]
Big data and artificial intelligence analytics in geosciences	[11]
Application review of big data and AI in geology	[12]
Game theory-based big data analytics in geosciences	[13]

final stage using sensitivity analysis. After such rigorous research, geologic features that were previously not identified (or interpreted) from the seismic data were revealed for even better decision-making.

Another research by Joshi et al. [19] worked upon mitigating hazardous situations. For instance, they used big data to analyse micro-seismic data for modelling fracture propagation maps during hydraulic fracturing in real time for its success (without any hassle or failures). They also combined exploration, drilling and production data to characterize reservoirs. Furthermore, their work also improved success ratio of operations by potential anomaly detection using historical datasets on success and failure(s).

A different study by Olneva et al. [20] used big data for clustering 1D, 2D & 3D geological maps for gaining insights on potential reservoirs and detecting geological patterns. They used seismic data and followed two different approaches named (by authors) as "from general to particulars" and "from particulars to general", for which they analysed patterns in 5000 wells in the first approach and 40,000 km^2 of area in the second approach.

The scenario of recent upsurge in the amount of research work upon the drilling operation in the industry is quite similar to the one discussed in the exploration part. For instance, Rijmenam focused on mitigating major environmental impacts while the real-time success prediction of drilling operations using 2D/3D/4D geological maps (with the help of multiple parallel processing big data platforms) was also mentioned [14,21].

This was again asserted by Soofi and Perez [16] by mentioning that big data can help in beforehand anomaly detection that might save millions in equipment and labour costs. They also mentioned that the use of real-time big data from Supervisory Control and Data Acquisition systems can vastly help in maximizing asset (such as equipment) performance and optimizing production. In fact, a mere 1% production increase from the current rate can literally add 2 years to the world's petroleum supply [22].

A major part of research is taken by research work that's focused on optimization of the drilling operations. This is in resonance with the battle of mitigating 10% average downtime in the industry (three times higher than other industries [23]). This can be reduced with the help of big data analytics by resolving possibilities of unanticipated equipment failure with actions such as alerting crews to repair or replace equipment before the point of failure. It can also be used for predicting possible maintenance requirements by constant monitoring and analysis (and thus reducing overall downtime) [24]. Moreover, combining the sensory data with geological data can also help in predicting failure as well as figuring out which equipment works best in a given environmental condition [25].

A study by Blair [26] stated the fact that big data analysis pertaining to geology and operations can drastically improve drilling, spacing of wells and well completion, despite ever-changing climate and environment (as the constant analysis can rectify and overcome such scenarios). Its usefulness was also mentioned in the drilling and finishing process of wells.

In a study by Duffy [27], the drilling rig efficiency was drastically improved by incorporating safe practices identified using an "automated drilling state detection monitoring service". In their case study on Bakken pad drilling, their results provided savings of around 12 days along with the total non-drilling time reduction by observed 45% (almost reduction by half). In another study by Maidla [28], data from electronic drilling recorder, differential pressure, morning report and cross-plots of weight on bit was utilized along with emphasis on data filtering, quality control and know-how of basic physics as critical factors for obtaining optimized outcome in drilling performance improvement.

Furthermore, there's quite a bit of research into reducing the risks associated with the drilling operation as well. A study by Johnston and Guichard [29] used drilling data, well logging and geological formation data (of about 350 oil and gas wells) in different formats (such as .txt, .xls, .pdf & .las) to form a big data set that can be analysed to optimize and reduce any drilling-related risks in the operations.

In a nutshell, the research work pertaining to big data's implementation in exploration and drilling has exploded and seems to keep going for at least the next decade when big data might become petroleum industry's norm. Moreover, this is in accordance with the fact that the analysis performed upon the big data can help in reservoir discoveries as well as measuring drilling operation success beforehand (through weather, soil and equipment data under seismic data) that directly affects the expenditure and revenue [25]. Besides, it also helps in reducing lag time, decreasing technical risks and improving drilling parameters (before it even begins) [30]. Moreover, the research work mentioned until now in this section is summarized in Table 9.3.

9.5   RESEARCH OF BIG DATA IN RESERVOIR STUDIES

Research by Bello et al. [31] employed big data for the creation of a reservoir management application that boosted performance optimization and improved the site management. Their new system provided several technical and economic values to O&G companies such as agile-based quick deployment, improved data quality, dynamic scalability, industry-driven development, web & mobile access, alarm notifications and robust & effective data management system operations.

Table 9.3 Research work pertaining to big data in exploration and drilling

Research	References
Role of supercomputers in seismic research	[14]
Identification of "productive seismic trace signatures"	[15]
Identification of reservoir traces	[16]
Identification of hydrocarbon deposits through 3D seismic data	[17]
Multi-component seismic analysis using PCA and SOMs	[18]
Mitigation of hazardous situations through micro-seismic data	[19]
Clustering of ID, 2D & 3D geological maps for gaining insights	[20]
Real-time success prediction of drilling operations using 2D/3D/4D geological maps	[14,21]
Beforehand anomaly detection	[16]
Resolving of unanticipated equipment failure possibilities and predicting of possible maintenance requirements	[24]
Prediction of failure and finding of best working environmental condition for equipment	[25]
Improvement of drilling, spacing of wells and well completion using big data analytics	[26]
"Automated drilling state detection monitoring service" for drilling rig efficiency improvement	[27]
Optimized outcome in drilling performance improvement using several channels/streams of big data	[28]
Optimization and reduction of drilling-related risks	[29]

Another research by Brulé [32] proposed closed-loop reservoir management and integrated asset modelling for managing reservoirs, under a term called "data reservoir", which utilized data management and analytics in exploration and production with the help of big data technologies for handling data of any volume as well as variety. Their proposal supports physics-based and empirical-based approaches for several services like real-time analytical processing, traditional BI & reporting and discovery analytics (for all forms of data), while removing limitations of relations databases in terms of scale.

There has been research into reducing the environmental impact of reservoir engineering as well. A proposal by Haghighat et al. [33] used big data and smart field technology for detection of CO_2 leakage in storage project and thus improvement in CO_2 sequestration. Their project developed the next generation of intelligent software that exploited the massive amount of big data to autonomously and continuously monitor as well as verify CO_2 sequestration in geological formations, with the help of artificial intelligence and data mining tools.

Moreover, as optimization has been at the core of big data's implementation, research has also focused on this aspect in reservoir studies. Research by Popa et al. [34] acknowledge the increased level of instrumental usage in heavy oil reservoirs (on top of producers, injectors and observation walls). This led them to propose a set of practical examples that, upon use of big data and analytics, can improve reservoir management, better decision-making process and enhance their operational performance.

A different research work in reservoir engineering has been into modelling improvement of hydraulically fractured reservoirs by Udegbe et al. [35], where they employed big data analytics for restimulation candidate selection. Their production data classification approach (adapted from real-time face detection) has shown prospects for long duration production data analysis for the said hydraulically fractured reservoirs, while possessing potentials for other applications as well due to its ease of modification.

Recollecting the advanced strategies of recovery of hydrocarbons in depleted or underperforming reservoirs, research work for big data in Enhanced Oil Recovery (EOR) has been on the rise. For instance, a study by Xiao and Sun [36] portrayed the significance of big data analytics in EOR projects for various benefits such as accurate reservoir dynamics prediction, reservoir simulation, different systems connectivity and continuous monitoring, while improving the optimization of implementation of EOR into O&G projects.

In a nutshell, the research work pertaining to big data's implementation in reservoir engineering has exploded as well and, due to its catered benefits, seems to keep going for at least the next few decades when big data might become petroleum industry's norm. Moreover, the research work mentioned until now in this section is summarized in Table 9.4.

9.6 RESEARCH OF BIG DATA IN PRODUCTION AND TRANSPORTATION

Big data has been employed in automated conduction of decline analysis (by Seemann at al. [37]) during production operations with the help of a "clustering, data mining and automation-based Decline Curve Analysis Clustering". It uses big data–enabled data mining for clustering purposes that in turn helps in automatic decline analysis.

Table 9.4 Research work pertaining to big data in reservoir studies

Research	References
Reservoir management application	[21]
Data reservoir: closed-loop reservoir management and integrated asset modelling	[32]
Detection of CO_2 leakage and improvement in CO_2 sequestration	[33]
Heavy oil reservoir optimization	[34]
Improved modelling of hydraulically fractured reservoirs	[35]
Big data analytics based optimization of EOR project applications	[36]

Another study by Rollins et al. [38] was focused on production allocation and analysis based on big data for production data. This allocation was a slow and complicated challenge that was required to be solved and hence the study was undertaken. They devised a mapping tool for petroleum engineers, equipping them to interface with the allocated production data while visualizing the regional production outcomes (or results) through formation. This leap forward provided a necessary research push into the management and allocation of production-based big data which is generally overlooked.

Due to low oil prices, there has been a push into research work on fracturing projects as well. Big data has grown increasingly important due to slashed oil prices, and hence a study by Betz [39] was focused upon improvements in hydraulic fracturing projects. In the study, various underground sensors were deployed for gaining valuable data including distributed borehole temperature data, toe and heel pressure data, micro-seismic data and uninterrupted production logging with fiber-optic line data. This big data was then passed through big data analytics for valuable insights. The outcomes of the study provided a more concrete and reliable form of information (from one or more pilot/wells), for making effective and constrictive decisions.

Research work has also been into the optimization of various aspects of production engineering. For instance, research work by Palmer and Turland [40] aimed at optimization of proactive rod pumps by leveraging big data for acceleration and improvement of operations. They create a visualization tool termed "Rod Pump Optimization Tool", which resulted in effective communication, time savings, operation performance improvements, better knowledge management and scalability.

Continuing with optimization work, a study by Gupta et al. [41] used big data analytics for creating a workflow to safeguard Electric submersible pumps (ESPs) in real time. Their workflow provided assistance to petroleum engineers in several aspects such as figuring out of trends & patterns in ESP operation, diagnosis of problems pertaining to ESPs and undertaking of remedial actions for monitoring and safeguard purposes. Another similar study by Sarapulov and Khabibullin [42] aimed to improve the operational efficiency of ESPs by application of big data tools for (majorly) unstructured data analysis. Their application of big data was successful and was able to show insights on decision-making process that was previously not possible.

In a nutshell, the research work pertaining to big data's implementation in transportation and production has also been on the rise and, due to its catered benefits, seems to keep going for at least the next few decades when big data might become petroleum industry's norm. Moreover, the research work mentioned until now in this section is summarized in Table 9.5.

Table 9.5 Research work pertaining to big data in production and transportation

Research	References
Automated Conduction of Decline Analysis	[37]
Big Data-Enabled Production Allocation and Analysis	[38]
Improvement of hydraulic fracturing projects	[39]
Optimization of ESPs	[41,42]
Optimization of proactive rod pumps' performance	[40]

9.7 RESEARCH OF BIG DATA IN PETROLEUM REFINERY

A research work by Khvostichenko and Makarychev-Mikhailov [43] studied the effect(s) of fracturing chemical additives on well productivity by conducting a statistical and rigorous assessment of the effects, based on dataset obtained from 48 US states. They essentially designed a workflow to study well completion parameters' impact on well productivity and, upon evaluation, came to a conclusion that while the use of gelling agents has no impact on the improvement of refinery productivity, flowback aids' use did provide a production improvement of 20%–60% which showed the existence of said impacts [43].

Another study by Anderson [44] was based on creation of Petroleum Analytics Learning Machine for optimization of Internet of Things (IoTs) in today's digital field-to-refinery petroleum system. During the experimentation, they utilized all petroleum attributes to compute importance weights while predicting petroleum and water production (which allows the prediction of accurate Estimated Ultimate Recovery over the lifetime of each production well). This helps tremendously comprehension and maintenance of various aspects of petroleum operations including refinery scheduling plans [44].

Fidelis and Effiong [45] conducted a holistic look on the improvements of effective utilization of big data in refining operations (by understanding inefficiency sources such as waste heat, steam systems & furnaces, delayed coker, reformers and distillation units) as they produce massive amount of data that is stored in ineffective and disengaged information systems. This data could be (and hence was) utilized to build data models that would aid in the reliability of operations. They adopted a research methodology that aims to mitigate factors in visualizing refining time-series data in a timely manner, in order to enforce effective refining operations [45].

In a nutshell, the research work pertaining to big data's implementation in petroleum engineering has exploded as well and, due to its catered benefits, seems to keep going for at least the next few decades when big data might become petroleum industry's norm. Moreover, the research work mentioned until now in this section is summarized in Table 9.6.

9.8 RESEARCH OF BIG DATA IN NATURAL GAS

Rijmenam [21] presented the opportunities offered by big data in maximizing the output of gas generation while mitigating major environmental impacts. It was also discussed that big data enables the discovery of anomalies beforehand which can help in minimizing downtimes as well as major environmental disasters.

Table 9.6 Research work pertaining to big data in petroleum refinery

Research	References
Workflow to study well completion parameters' impact on well productivity	[43]
Petroleum Analytics Learning Machine-based IoT optimization for field-to-refinery O&G system	[44]
Utilization of Big Data Analytics for Effective Refinery Operations	[45]

The research has also been into ensuring safety and reduction of hazards in the petroleum workspace (especially extraction operations in natural gas). Drilling sensory data along with weather data can be meticulously used to predict and avoid dangerous situations for workers and potentially mitigate any related environmental risks, as noted by Soofi et al. [16,25].

Even natural calamities such as earthquakes can be sensed through constant big data stream gathered from sensors in real time to predict whether a shutdown of facility would be required (or safety measures be taken into account) for everyone's safety, as reported by Blair [26].

In a nutshell, the research work pertaining to big data's implementation in petroleum natural gas has exploded and, due to its catered benefits, seems to keep going for at least the next few decades when big data might become petroleum industry's norm. Moreover, the research work mentioned until now in this section is summarized in Table 9.7.

9.9   RESEARCH OF BIG DATA IN HEALTH AND SAFETY

For instance, a combination of research work undertaken by Tarrahi and Shadravan [46,47] was focused on increasing the occupational safety of the O&G industry [46,48]. In the first study by Tarrahi and Shadravan [47], they accepted the fact that most conventional HSE (Health, Safety and Environment) management and hazard identification systems do not possess the capability for agile and automated data integration and smart decision-making. Thus, they presented the use of big data analytics on humongous dataset to extract exploratory features' relationship for making intelligent decision strategies for HSE management. Their experimentation on a public dataset with big data analytics algorithm was able to explore 3 million records (divided into categories such as datatype, case type, area and year among others) for an improved safety leadership [47].

Moreover, the other study by Tarrahi and Shadravan [46] was focused on intelligent HSE big data analytics platform for the promotion of occupational safety, due to the

Table 9.7 Research work pertaining to big data in petroleum natural gas

Research	References
Predicting and avoiding dangerous situations and mitigating environmental risks using drilling sensory data	[16,25]
Natural calamity sensing for facility shutdown prediction	[26]
Maximization of oil and gas output generation along with major environmental impacts	[21]

aforementioned fact about the inability of conventional systems to perform agile and automated data integration-based tasks. Their experimentation was based on Bureau of Labour Statistics dataset with 1278 industries, 18 natures of incident, 846 sources of injury and 18 worker locations between 2011 and 2014. Their outcomes showed the promotion and improvement of safety in various petroleum operations [46].

Another study by Park et al. [48]) proposed a model for calculation of Energy Efficiency Operational Indicator using only public data (without actual operating data) which are measured during ship operation. The experimental public data contained Automatic Identification System data and marine environment data, which would be tremendously difficult to analyse with the existing data processing method (that has poor energy efficiency). Therefore, big data technology was studied and applied by the researchers for utilizing the inherent capability of public data to provide insights effectively. The application of the proposed model on the example ship was done for estimation of the ship energy efficiency indicator and the effectiveness of the model. The analysis outcomes showed improvement of Energy Efficiency Operational Indicator's calculation using big data with viable effects on health and safety of operations [48].

Pettinger [49] described the role of leading indicators, termed as "golden eggs", in the safety culture of petroleum industry. The search of safety professionals for true leading indicators of a great safety culture has generally been met with frustrated identifying, tracking and acting upon these indicators. Therefore, the paper presented the benefits of using big data-based safety predictive analytics for the identification of the leading indicators in order to eliminate death in the workplace [49].

A different research work by Cadei et al. [50] used H_2S concentration for the prediction of hazardous event(s) by developing a prediction software to forecast operational upsets during petroleum production and hazard events. The experimentation of the proposed software was performed upon a dataset containing data from different sources (such as real-time stories, historical data, operator data and maintenance reports) which was processed by modelling (with the help of artificial neural network) and model validation. Their analysis outcomes showed that it's possible to quickly predict hazardous events and accordingly figure out feasible asset configuration for improved health and safety [50].

Another study by Ajayi et al. [51] developed a health hazard analytics model comprising six stages for improved decision-making using better hazard big data analytics. The stages included first stage of data preparation (for identification and fixation of errors within the data). The second one was exploratory analytics and model selection (where a suitable analytical model was picked for particular datasets). In the third stage, analytical models were designed for health and safety identification & evaluation. This was followed by parameter extraction and model execution in the fourth stage. Thereafter, health forecasting was undertaken is the fifth stage while prescriptive analysis was done in the sixth stage. The proposed model, upon experimentation, was able to provide data analytic outcomes for better decision-making [51].

Tanabe et al. [52] proposed a big data-enabled safety model design in their study for onshore natural gas liquefaction plant, which consisted of the "safety critical design basis matrix" for better and accurate identification of external common cause(s) of failure. Upon evaluation, the model provided shield to liquefied natural gas cryogenic spill, while maximizing natural air circulation (for prevention of plant accidents) [52]. Another study by Xie et al. [53] proposed a structure framework for factor detection that influences and predicts the failure rate of safety instrumented systems equipment. The proposed framework comprised statistical and data-driven models (such as PCA and PLSR) for failure rates' prediction [53].

In a nutshell, the research work pertaining to big data's implementation in petroleum health and safety has exploded as well and, due to its catered benefits, seems to keep going for at least the next few decades when big data might become petroleum industry's norm. Moreover, the research work mentioned until now in this section is summarized in Table 9.8.

9.10 RESEARCH OF BIG DATA IN FINANCE AND FINANCIAL MARKETS

The wake of technological advancements has brought in a significant amount of research work pertaining to big data in the last decade, and into the field of petroleum finance and financial markets as well. A study by Yu et al. [54] was aimed at understanding Google trends for accurate oil consumption prediction which could potentially help the petroleum industry with production planning and product distribution management, resulting in massive economic upside. They used the rich mine of online Google trend big data and fed it to a model with two main steps, relationship investigation and prediction improvement. The experimental results clearly portrayed significant performance improvements in both directional and level predictions [54].

Another study by Duan et al. [55] was focused on big data-enabled dynamic risk assessment of the Overseas O&G investment environment by revealing the specific occurrence probabilities of risk on different levels. Their research selected 25 indicators from six dimensions and applied a Bayesian network algorithm to dynamically assess the O&G overseas investment risk of ten countries. The experimental results revealed how the risk dynamics have changed over the past two decades, while serving as a reference in future overseas O&G investment risk decision-making processes (playing a significant role in outbound investing, engineering and service projects) [55].

A different study by Popovič et al. [56] was dedicated to understanding of the impact that big data analytics can have on any organization's high value business performance. Their research findings indicate that big data analytics' capability of data

Table 9.8 Research work pertaining to big data in petroleum health and safety

Research	References
Energy efficiency model development for ship operations	[48]
HSE management improvement by advanced big data analytics	[47]
Occupational Safety promotion with Intelligent HSE big data analytics platform	[46]
Safety predictive analytics development for death elimination	[49]
Hazard events forecast software development	[50]
Big data platform-based health and safety accident prediction	[51]
Big data analytics-based safety design approach for onshore LNG liquefaction plant	[52]
Operational data-driven prediction for safety instrumented systems equipment failure rates	[53]

sourcing, access, integration and delivery, analytical strengths and people's expertise along with organizational readiness and design factors (such as big data strategy, top management support, financial resources and employee engagement) can result in better financial decision-making and planning, and thus provide enhanced high value business performance [56].

A study by Żbikowski [57] was focused on time series prediction and recommended a modified version of Support Vector Machine (SVM) classifier, namely volume-weighted SVM (VW-SVM) for the same. The idea behind the proposition was based on the hypothesis that integrating transaction volume into the loss function can provide better performance future trend prediction by the classifier (which could further be optimized using Sequential Minimal Optimization algorithm). The research data used for experimentation comprised daily quotations of 420 stocks present in the S&P500 stock index. Moreover, various technical indicators were utilized as input, and two classes were selected that indicated short-term trends (upward and downward) for target value. The experimental results demonstrate that VW-SVM can effectively forecast short-term trends in stock price movement (up or down), and beat traditional algorithms or models in terms of performance [57].

Yu and Yan [58] approached the prediction and analysis of the financial data as a time-dependent and linear problem, for which a Deep Neural Network (DNN)-based prediction model built upon the phase-space reconstruction method and Long Short-Term Memory (LSTM) network. The prediction was performed in three stages, namely, time-series data processing, architecture building and result evaluation & analysis. The complete analysis was carried out upon indices like S&P500, Dow Jones Industrial Average (DJIA), Hang Seng index (HSI), CSI300, Nikkei225 and ChiNext, and against models like AutoRegressive Integrated Moving Average (ARIMA) linear analytical model, Support Vector Regression (SVR) ML model and a deep MLP model. The results (with Tanh activation function, as determined after comparison between Tanh and rectified linear unit/ReLu) show maximum accuracy reached by LSTM to be 62.87% (for S&P500) which was clearly greater than the compared models. Thus, they concluded that the established DNN-based LSTM model portrays a higher capability for predicting time series and is a viable option for use in trading strategies [58].

Chandak [59] approached the stock market prediction using the LSTM model with two main layers where each layer consisted of 50 nodes (with standard sigmoid activation function). They used the finance data of 5 years from Yahoo Finance with 1259 rows (out of which 1000 rows were utilized for training dataset and the rest 259 rows for testing dataset). With number of epochs set to 1, a loss of 0.0018 (mean squared error) was achieved with 59 seconds of training time, while setting number of epochs to 5 resulted in a loss of 0.00032676 within 280 seconds. It was noted that as the number of nodes was increased, the accuracy went up, but so did the computational resources as well as time required for training. With the promising results, the study clearly established the effectiveness of LSTMs in time series data prediction [59].

In a nutshell, the research work pertaining to big data's implementation in petroleum finance and financial markets has exploded as well and, due to its catered benefits, seems to keep going for at least the next few decades when big data might become petroleum industry's norm. Moreover, the research work mentioned until now in this section is summarized in Table 9.9.

Table 9.9 Research work pertaining to big data in petroleum finance and financial markets

Research	References
Google trends big data-based oil consumption forecasting	[54]
Big data-based dynamic risk assessment of the overseas O&G investment environment	[55]
Big data analytics-based high value business performance	[56]
VW-SVM-based time series prediction	[57]
LSTM-based time series prediction	[59]
DNN-based LSTM time series prediction	[58]

9.11 CONCLUSION

As extensively discussed throughout the chapter, there has been explosion in the amount of research work in diverse operations of the oil and gas industry that helps in future applications that provide massive upsides to its implementation (including reduction of costs, decrease in time, improvement in efficiency, decrease in environmental impact and ensured upholding of regulations) among various others.

Some of the futuristic research work is already in implementation in the real world, which happens to be the topic of discussion in the next chapter on real world implementation of big data. Exploiting the best of the capabilities of big data, O&G companies can see drastic improvements which would else seem impossible to achieve.

REFERENCES

1. von Plate, M. (2016). Big Data Analytics for Prognostic Foresight. *SPE Intelligent Energy International Conference and Exhibition 2016*, 1–6. https://doi.org/10.2118/181037-ms.
2. Anagnostopoulos, A. (2018). Big Data Techniques for Ship Performance Study. *Proceedings of the 28th International Ocean and Polar Engineering Conference*, 887–893. https://onepetro.org/ISOPEIOPEC/proceedings-abstract/ISOPE18/All-ISOPE18/ISOPE-I-18-190/20389.
3. Yin, Q., Yang, J., Zhou, B., Jiang, M., Chen, X., Fu, C., Yan, L., Li, L., Li, Y., & Liu, Z. (2018). Improve the Drilling Operations Efficiency by the Big Data Mining of Real-Time Logging. *SPE/IADC Middle East Drilling Technology Conference and Exhibition 2018*, 1–12. https://doi.org/10.2118/189330-ms.
4. Azadeh, A., Ghaderi, S., & Asadzadeh, S. (2008). Energy efficiency modelling and estimation in petroleum refining industry – A comparison using physical data. *Renewable Energy and Power Quality Journal*, 1(6), 123–128. https://doi.org/10.24084/repqj06.242.
5. Wang, T., Li, T., Xia, Y., Zhang, Z., & Jin, S. (2017). Risk assessment and online forewarning of oil & gas storage and transportation facilities based on data mining. *Procedia Computer Science*, 112, 1945–1953. https://doi.org/10.1016/j.procs.2017.08.052.
6. Fan, M. W., Ao, C. C., & Wang, X. R. (2019). Comprehensive method of natural gas pipeline efficiency evaluation based on energy and big data analysis. *Energy*, 188, 1–12. https://doi.org/10.1016/j.energy.2019.116069.
7. Tokarek, T. W., Odame-Ankrah, C. A., Huo, J. A., McLaren, R., Lee, A. K. Y., Adam, M. G., Willis, M. D., Abbatt, J. P. D., Mihele, C., Darlington, A., Mittermeier, R. L., Strawbridge, K., Hayden, K. L., Olfert, J. S., Schnitzler, E. G., Brownsey, D. K., Assad, F. V., Wentworth, G. R., Tevlin, A. G., . . . Osthoff, H. D. (2018). Principal

component analysis of summertime ground site measurements in the Athabasca oil sands with a focus on analytically unresolved intermediate-volatility organic compounds. *Atmospheric Chemistry and Physics*, 18(24), 17819–17841. https://doi.org/10.5194/acp-18-17819-2018.

8. Lin, A. (2014). Principles of Big Data Algorithms and Application for Unconventional Oil and Gas Resources. *SPE Large Scale Computing and Big Data Challenges in Reservoir Simulation Conference and Exhibition 2014*, 1–9. https://doi.org/10.2118/172982-ms.

9. Chelmis, C., Zhao, J., Sorathia, V., Agarwal, S., & Prasanna, V. (2012). Semiautomatic, Semantic Assistance to Manual Curation of Data in Smart Oil Fields. *SPE Western Region Meeting 2012*, 1–18. https://doi.org/10.2118/153271-ms.

10. Qi, Z., & Xuelong, L. (2019). Big data: New methods and ideas in geological scientific research. *Big Earth Data*, 3(1), 1–7. https://doi.org/10.1080/20964471.2018.1564478.

11. Spina, R. (2018). Big data and artificial intelligence analytics in geosciences: Promises and potential. *GSA Today*, 29(1), 42–43. https://www.geosociety.org/gsatoday/groundwork/G372GW/article.htm.

12. Chen, L., Wang, L., Miao, J., Gao, H., Zhang, Y., Yao, Y., Bai, M., Mei, L., & He, J. (2020). Review of the application of big data and artificial intelligence in geology. *Journal of Physics: Conference Series*, 1684, 1–6. https://doi.org/10.1088/1742-6596/1684/1/012007.

13. Bruce, L. M. (2013). Game Theory Applied to Big Data Analytics in Geosciences and Remote Sensing. *2013 IEEE International Geoscience and Remote Sensing Symposium – IGARSS*, 4094–4097. https://doi.org/10.1109/igarss.2013.6723733.

14. Marr, B. (2015). Big Data in Big Oil: How Shell Uses Analytics to Drive Business Success. https://www.forbes.com/sites/bernardmarr/2015/05/26/big-data-in-big-oil-how-shell-uses-analytics-to-drive-business-success/.

15. Feblowitz, J. (2013). Analytics in Oil and Gas: The Big Deal about Big Data. *SPE Digital Energy Conference and Exhibition 2013*, 1–6. https://doi.org/10.2118/163717-ms.

16. Soofi, A., & Perez, E. (2014). Drilling for New Business Value: How Innovative Oil and Gas Companies are Using Big Data to Outmaneuver the Competition. https://cloudblogs.microsoft.com/industry-blog/uncategorized/2015/10/18/drilling-new-business-value-innovative-oil-gas-companies-using-big-data-maneuver-competition/.

17. Baaziz, A, & Quoniam, L. (2013). How to use big data technologies to optimize operations in upstream petroleum industry. *International Journal of Innovation*, 1(1), 19–25.

18. Roden, R. (2016). Seismic Interpretation in the Age of Big Data. SEG Technical Program Expanded Abstracts 2016, 4911–4915. https://doi.org/10.1190/segam2016-13612308.1.

19. Joshi, P., Thapliyal, R., Chittambakkam, A. A., Ghosh, R., Bhowmick, S., & Khan, S. N. (2018). Big Data Analytics for Micro-Seismic Monitoring. *Offshore Technology Conference Asia 2018*, 1–5. https://doi.org/10.4043/28381-ms.

20. Olneva, T., Kuzmin, D., Rasskazova, S., & Timirgalin, A. (2018). Big Data Approach for Geological Study of the Big Region West Siberia. *SPE Annual Technical Conference and Exhibition 2018*, 1–6. https://doi.org/10.2118/191726-ms.

21. Rijmenam, M. (2015). If Big Data Is the New Oil, the Oil and Gas Industry Knows How to Handle It. https://datafloq.com/read/big-data-oil-oil-gas-industry-handle/133.

22. Cowles, D. (2015). Oil, Gas, and Data: High-Performance Data Tools in the Production of Industrial Power. https://www.oreilly.com/content/oil-gas-data/.

23. Dodgson, L. (2016). Cutting Costs: How Big Data Can Help. https://www.offshore-technology.com/features/featurecutting-costs-how-big-data-can-help-4733715/.

24. Harpham, B. (2016). How Data Science Is Changing the Energy Industry. https://www.cio.com/article/3052934/how-data-science-is-changing-the-energy-industry.html.

25. Natkar, S. (2016). Big Value for Big Data in Oil and Gas Industry! https://www.esds.co.in/blog/big-value-big-data-oil-gas-industry/.

26. Hassani, H., & Silva, E. S. (2018). Big data: A big opportunity for the petroleum and petrochemical industry. *OPEC Energy Review*, 42(1), 74–89. https://ualresearchonline.arts.ac.uk/id/eprint/13142/.

27. Duffy, W., Rigg, J., & Maidla, E. (2017). Efficiency Improvement in the Bakken Realized through Drilling Data Processing Automation and the Recognition and Standardization of Best Safe Practices. *SPE/IADC Drilling Conference and Exhibition 2017*, 1–10. https://doi.org/10.2118/184724-ms.

28. Maidla, E., Maidla, W., Rigg, J., Crumrine, M., & Wolf-Zoellner, P. (2018). Drilling Analysis Using Big Data has been Misused and Abused. *SPE/IADC Drilling Conference and Exhibition 2018*, 1–25. https://doi.org/10.2118/189583-ms.

29. Johnston, J., & Guichard, A. (2015). New Findings in Drilling and Wells Using Big Data Analytics. *Offshore Technology Conference 2015*, 1–8. https://doi.org/10.4043/26021-ms.

30. Technavio (2015). How Oil and Gas Is Using Big Data for Better Operations. https://blog.technavio.com/blog/how-oil-and-gas-using-big-data-better-operations.

31. Bello, O., Yang, D., Lazarus, S., Wang, X. S., & Denney, T. (2017). Next Generation Downhole Big Data Platform for Dynamic Data-Driven Well and Reservoir Management. *SPE Reservoir Characterisation and Simulation Conference and Exhibition 2017*, 1–38. https://doi.org/10.2118/186033-ms.

32. Brulé, M. (2015). The Data Reservoir: How Big Data Technologies Advance Data Management and Analytics in E&P. *SPE Digital Energy Conference and Exhibition 2015*, 1–7. https://doi.org/10.2118/173445-ms.

33. Haghighat, S. A., Mohaghegh, S. D., Gholami, V., Shahkarami, A., & Moreno, D. (2013). Using Big Data and Smart Field Technology for Detecting Leakage in a CO_2 Storage Project. *SPE Annual Technical Conference and Exhibition 2013*, 1–7. https://doi.org/10.2118/166137-ms.

34. Popa, A. S., Grijalva, E., Cassidy, S., Medel, J., & Cover, A. (2015). Intelligent Use of Big Data for Heavy Oil Reservoir Management. *SPE Annual Technical Conference and Exhibition 2015*, 1–14. https://doi.org/10.2118/174912-ms.

35. Udegbe, E., Morgan, E., & Srinivasan, S. (2017). From Face Detection to Fractured Reservoir Characterization: Big Data Analytics for Restimulation Candidate Selection. *SPE Annual Technical Conference and Exhibition 2017*, 1–20. https://doi.org/10.2118/187328-ms.

36. Xiao, J., & Sun, X. (2017). Big Data Analytics Drive EOR Projects. *SPE Offshore Europe Conference & Exhibition 2017*, 1–9. https://doi.org/10.2118/186159-ms.

37. Seemann, D., Williamson, M., & Hasan, S. (2013). Improving Reservoir Management through Big Data Technologies. *SPE Middle East Intelligent Energy Conference and Exhibition 2013*, 1–11. https://doi.org/10.2118/167482-ms.

38. Rollins, B. T., Broussard, A., Cummins, B., Smiley, A., & Eason, T. (2017). Continental Production Allocation and Analysis through Big Data. *Proceedings of the 5th Unconventional Resources Technology Conference*, 2053–2060. https://doi.org/10.15530/urtec-2017-2678296.

39. Betz, J. (2015). Low oil prices increase value of big data in fracturing. *Journal of Petroleum Technology*, 67(4), 60–61. https://doi.org/10.2118/0415-0060-jpt.

40. Palmer, T., & Turland, M. (2016). Proactive Rod Pump Optimization: Leveraging Big Data to Accelerate and Improve Operations. *SPE North America Artificial Lift Conference and Exhibition 2016*, 1–15. https://doi.org/10.2118/181216-ms.

41. Gupta, S., Saputelli, L., & Nikolaou, M. (2016). Big Data Analytics Workflow to Safeguard ESP Operations in Real-Time. *SPE North America Artificial Lift Conference and Exhibition 2016*, 1–14. https://doi.org/10.2118/181224-ms.

42. Sarapulov, N. P., & Khabibullin, R. A. (2017). Application of Big Data Tools for Unstructured Data Analysis to Improve ESP Operation Efficiency. *SPE Russian Petroleum Technology Conference 2017*, 1–11. https://doi.org/10.2118/187738-ms.

43. Khvostichenko, D., & Makarychev-Mikhailov, S. (2018). Effect of Fracturing Chemicals on Well Productivity: Avoiding Pitfalls in Big Data Analysis. *SPE International Conference and Exhibition on Formation Damage Control 2018*, 1–13. https://doi.org/10.2118/189551-ms.

44. Anderson, R. N. (2017). "Petroleum Analytics Learning Machine" for Optimizing the Internet of Things of Today's Digital Oil Field-to-Refinery Petroleum System. *2017 IEEE International Conference on Big Data (Big Data)*, 4542–4545. https://doi.org/10.1109/bigdata. 2017.8258496.

45. Fidelis, E. W., & Effiong, A. (2019). Utilization of Big Data Analytics for Effective Refinery Operations. *NAPE Annual International Conference 2019*, 29(1), 113–120. https://nape.org.ng/wp-content/uploads/2020/09/113-120.-Utilization-of-Big-Data-Analysis.pdf.

46. Tarrahi, M., & Shadravan, A. (2016b). Intelligent HSE Big Data Analytics Platform Promotes Occupational Safety. *SPE Annual Technical Conference and Exhibition 2016*, 1–21. https://doi.org/10.2118/181730-ms.

47. Tarrahi, M., & Shadravan, A. (2016a). Advanced Big Data Analytics Improves HSE Management. *SPE Bergen One Day Seminar 2016*, 1–7. https://doi.org/10.2118/180032-ms.

48. Park, S. W., Roh, M. I., Oh, M. J., Kim, S. H., Lee, W. J., Kim, I. I., & Kim, C. Y. (2018). Estimation Model of Energy Efficiency Operational Indicator Using Public Data Based on Big Data Technology. *The 28th International Ocean and Polar Engineering Conference*, Sapporo, Japan.

49. Pettinger, C. B. (2014). Leading Indicators, Culture and Big Data: Using Your Data to Eliminate Death. *ASSE Professional Development Conference and Exposition 2014*, Orlando, FL.

50. Cadei, L., Montini, M., Landi, F., Porcelli, F., Michetti, V., Origgi, M., Tonegutti, M., & Duranton, S. (2018). Big Data Advanced Analytics to Forecast Operational Upsets in Upstream Production System. *Abu Dhabi International Petroleum Exhibition & Conference 2018*, 1–14. https://doi.org/10.2118/193190-ms.

51. Ajayi, A., Oyedele, L., Davila Delgado, J. M., Akanbi, L., Bilal, M., Akinade, O., & Olawale, O. (2019). Big data platform for health and safety accident prediction. *World Journal of Science, Technology and Sustainable Development*, 16(1), 2–21. https://doi.org/10.1108/wjstsd-05-2018-0042.

52. Tanabe, M., & Miyake, A. (2010). Safety design approach for onshore modularized LNG liquefaction plant. *Journal of Loss Prevention in the Process Industries*, 23(4), 507–514. https://doi.org/10.1016/j.jlp.2010.04.004.

53. Xie, L., Håbrekke, S., Liu, Y., & Lundteigen, M. A. (2019). Operational data-driven prediction for failure rates of equipment in safety instrumented systems: A case study from the oil and gas industry. *Journal of Loss Prevention in the Process Industries*, 60, 96–105. https://doi.org/10.1016/j.jlp.2019.04.004.

54. Yu, L., Zhao, Y., Tang, L., & Yang, Z. (2019). Online big data-driven oil consumption forecasting with Google trends. *International Journal of Forecasting*, 35(1), 213–223. https://doi.org/10.1016/j.ijforecast.2017.11.005.

55. Duan, X., Zhao, X., Liu, J., Zhang, S., & Luo, D. (2021). Dynamic risk assessment of the overseas oil and gas investment environment in the big data era. *Frontiers in Energy Research*, 9. https://doi.org/10.3389/fenrg.2021.638437.

56. Popovič, A., Hackney, R., Tassabehji, R., & Castelli, M. (2016). The impact of big data analytics on firms' high value business performance. *Information Systems Frontiers*, 20(2), 209–222. https://doi.org/10.1007/s10796-016-9720-4.

57. Żbikowski, K. (2014). Time Series Forecasting with Volume Weighted Support Vector Machines. *Communications in Computer and Information Science*, 250–258. https://doi.org/10.1007/978-3-319-06932-6_24.

58. Yu, P., & Yan, X. (2019). Stock price prediction based on deep neural networks. *Neural Computing and Applications*, 32(6), 1609–1628. https://doi.org/10.1007/s00521-019-04212-x.

59. Chandak, A. V. (2020). Stock market predictor using long short-term memory (LSTM) technique. *International Journal of Innovative Technology and Exploring Engineering*, 9(4), 393–396.https://doi.org/10.35940/ijitee.d9075.029420.

Chapter 10

Real-World Implementation of Big Data

10.1 INTRODUCTION

The success of significant contribution of the research community on advancements in big data research in each petroleum industry operation can be validated by assessing the level of real-world implementation undertaken by the Oil and Gas (O&G) companies in the real world. It aids in comprehending whether the benefits along with technological advancements brought forward by the research are actually worth in the economic and conservative world of oil and gas. Keeping this in mind, the outlook does seem positive as there has been significant amount of big data application by several companies around the world.

Few real-world implementations by O&G companies on big data include efficiency improvements, environmental impact reduction, seismic data analysis and refinery plant optimization. A deeper dive into all of them into the industry (and subsequently each operation) will begin with the next section. This chapter outlines the industry-wide implementations (Section 10.2), real-world implementation in Geoscience (Section 10.3), Exploration and Drilling (Section 10.4), Reservoir Studies (Section 10.5), Production and Transportation (Section 10.6), Petroleum Refinery (Section 10.7), Natural Gas (Section 10.8), Health and Safety (Section 10.9) and Finance and Financial Markets (Section 10.10), followed by a conclusion (Section 10.11).

10.2 INDUSTRY-WIDE REAL-WORLD IMPLEMENTATION

Following the research work, there has been real-world implementation of big data into industry-wide operations. For instance, Shell uses big data for reducing oil extraction and gas extraction costs using partnership with Hewlett-Packard and Amazon Web Services. The data that they collaboratively gather is in the range of millions of readings through fibre-optic cables and helps in gaining meaningful and stark accurate understanding of what lies where underneath the surface [1–3]. Moreover, Shell also utilizes big data generated from sensory equipment inside exploration activities to ensure lowest downtime and overhead costs that eventually result in positive economic upside(s) [2].

Other companies like British Petroleum and Chevron have used big data (and Wide azimuth towed streamer acquisition – WATS technology) for locating new reservoirs

DOI: 10.1201/9781003185710-10

(including conventional and unconventional on top of associated reservoirs) by generating high-resolution topographic maps (or reservoir n-dimensional models) of surface under the earth and beneath salt canopies [4].

Several companies are using big data breakthroughs in their operations as well. For instance, companies like Environmental BioTechnologies and Glori Energy are complementing seismic data with novel sensors for their benefit, while other companies like HiFi Engineering and Silixa are using hardware like connected tools and smart pumps for exploiting big data's potential [4,5].

Big data is also used in "not-so-common aspects" of oil and gas industry operations as well, i.e., aid to workers in the form of telemedical help, insurance and security. As an example, a company named NuPhysicia uses benefits of big data analysis to provide telemedicine support to their workers who are situated far from healthcare centres [6]. This helps in ensuring the health and safety of the individuals who handle industry operations and keep it running. Health and safety of workers is a major factor under supervision by companies (as discussed in Petroleum Operation chapter's section on Health and Safety), and thus big data's aid in any manner is highly appreciated.

Devon Energy, on the other hand, was able to determine the cause of non-productive time in their operations and mitigate them using Hadoop, SAS and text data, which in turn resulted in enormous reduction of 30% of their non-productive time [7,8]. Moreover, as noted by Kambouris [7], oil and gas companies can get between $500,000 and $1,000,000 per day in savings by simply reducing their high-impact non-productive time.

There has been big data implementation in management activities in various petroleum operations as well. For instance, Infosys (an Indian MNC) provides various services in the same domain such as reservoir characterization, production monitoring, operation modelling and simulation [9]. Also, another company named Halliburton (an oil & gas service company) used big data techniques for optimizing their seismic space and drilling space while attaining better well planning [4]. Moreover, all the aforementioned real-world implementations are summarized in Table 10.1.

Table 10.1 Industry-wide real-world implementation of big data

Companies	Implementation
Shell	Reduction in oil extraction costs by accurate understanding what lies where underneath the surface and ensuring lowest downtime and overhead costs
British Petroleum and Chevron	Locating new reservoirs with high-resolution topographic maps of surface under the earth & beneath salt canopies
BioTechnologies and Glori Energy	Complementing seismic data with novel sensors for economic benefit
HiFi Engineering and Silixa	Usage of hardware like connected tools & smart pumps for exploiting big data's potential for economic benefits
NuPhysicia	Providing telemedicine support to workers who are situated far from healthcare centres
Halliburton	Optimization of seismic and drilling space while attaining better well planning
Devon Energy	Determination and mitigation of cause for company's non-productive time in their operations
Infosys	Production and transportation services for O&G companies

10.3 REAL-WORLD IMPLEMENTATION OF BIG DATA IN GEOSCIENCE

There has been significant real-world implementation by O&G companies and governmental entities in the earth science space. For instance, an initiative named Earth-Server is a solution (led by a collection of scientific communities and international initiatives) that provides a holistic approach which ranges from query languages and scalability up to mobile access and visualization; and is coordinated by Jacobs University of Bremen, European research centres, private companies and NASA. The implementation aims for the development of specific solutions that support open access and ad-hoc analytics on GeoScience big data (based on the OGC geoservice standards) [10].

Another venture initiated by Central Water Commission (CWC, Ministry of Water Resources, Govt. of India) and Indian Space Research Organization (ISRO, Department of Space, Govt. of India) termed Water Resources Information System or WRIS aims to assimilate and help make the analysis process of the same easier for earth science implementations, where data collection, generation and presentation happen on consistent basis [11,12]. The latest versions of the web app also consist of spatial layers and attributes that help government and private entities to use the same for real-world applications in various scenarios.

A similar initiative by NASA termed "Climate@Home" [13] is focused on advancement of climate modelling-based geoscience studies by taking massive amount of big data into consideration [14,15]. For instance, a meteorological satellite named "Himawari-9" collects roughly 3 TB data from space every single day [16]. The platform has since been used to feed data to various real-world applications by O&G companies due to legitimacy and reliability. Moreover, all the aforementioned real-world implementations are summarized in Table 10.2.

10.4 REAL-WORLD IMPLEMENTATION OF BIG DATA IN EXPLORATION AND DRILLING

The industry has picked up on real-world implementation of big data into drilling operations as well due to its benefits, and also partially because of the fact that Bureau of Safety and Environmental Enforcement (BSEE) requires offshore O&G drillers to monitor their equipment in real time while archiving the generated data onshore, forcing companies to adopt big data due to consequences faced by the said regulations [17]. Moreover, given the

Table 10.2 Real-world implementation of big data in geoscience

Companies	Implementation
JUB, European research centres, NASA and private companies	Specific solutions that support open access and ad-hoc analytics on GeoScience big data
Indian Government	Assimilation, generation and presentation of big data (along with spatial layers and attributes) to make analysis easier for earth science implementations
NASA	Advancement of climate modelling–based geoscience studies

fact that just the drilling of a deep water well costs over $100 million [1], the companies do have major economic benefits that push them to utilize big data nonetheless.

Saudi Aramco uses big data analysis to develop the automatic system that monitors the well kick by adopting five big data machine learning (ML) algorithms – Decision Tree (DT), Bayesian Network, Sequential Minimum Optimization (SMO), Artificial Neural Network (ANN) and K-Nearest Neighbour (KNN). These models use data from ground, drilling and logging parameters for predicting the well kick while also predicting other aspects such as flow, pressure, torque, drilling time and pump speed with accuracy over 90% (for top performing DT and KNN).

Gazprom Neft, a Russian O&G company, uses Teradata ASTER system for analysing oil well operations within a framework of big data pilot program (completed in 2015) [18]. The company's implementation is based on research where 200 million logging data points and accident logs (recorded by 1649 wells in 2014) were taken for analysis whose result clearly portrayed that big data provided a new cognition beneficial to the company.

Furthermore, Schlumberger (a technology oilfield services company) happens to stay the leader in big data technology implementation in the drilling operation sector. It collaborates with Microsoft and Google among other internet-based companies to make use of cloud infrastructure for them to bring the safety and expansibility of data processing. The company employs 250 software engineers around the globe in software development centres, focusing on big data, high-performance computing, cloud services, data analytics, Internet of Things (IoT), visualization and user experience. Moreover, all the aforementioned real-world implementations are summarized in Table 10.3.

10.5 REAL-WORLD IMPLEMENTATION OF BIG DATA IN RESERVOIR STUDIES

There has been a rise in companies who provide services pertaining to reservoir engineering too. For instance, SGS (a European company) provides a variety of reservoir engineering services including numerical reservoir modelling, production forecasting, decline curve analysis, well-test design & interpretation, development planning, reservoir fluid analysis (PVT), well-design optimization and economic modelling [19].

Another company Flotek (a technology-driven with specialty in chemistry and data) that serves customers across industrial, commercial and consumer markets also utilized big data to transform reservoir management and recovery from reactive to predictive [20,21].

Table 10.3 Real-world implementation of big data in exploration and drilling

Companies	Implementation
Saudi Aramco	Big data analysis–based automatic system for monitoring the well kick using 5 ML algorithms
Gazprom Neft	Oil well operations analysis using teradata ASTER system within a framework of big data pilot program
Schlumberger	Big data technology implementation in drilling operation sector by providing cloud infrastructure of them to bring the safety and expansibility of data processing

Even Shell uses big data analytics to analyse data using artificial intelligence technologies to create 3D and 4D maps of the oil reservoirs to find out how much oil and gas is still left in them (with the help of optical fibre cables with sensors within the wells to measure seismic data), aiding in effective study of reservoir which can eventually help in effective decision-making process(es) [22]. Moreover, all the aforementioned real-world implementations are summarized in Table 10.4.

10.6   REAL-WORLD IMPLEMENTATION OF BIG DATA IN PRODUCTION AND TRANSPORTATION

Shell uses sensory data in other aspects as well, such as improvement of its performance while proactively understanding which equipment requires maintenance! The company has managed to save over $1 million by leveraging sensor data analytics in Nigeria. It also uses big data to analyse production and transportation costs and associated economic factors that help in determination of how and where to move refined products and set the prices [22].

The world's largest listed oilfield services group, Schlumberger, has launched a software named "Delfi" that aids in the maximization of output from an entire oilfield through improved coordination of production well design and drilling. It alone expects to reduce the production costs by 40% within the next decade [23].

General Electric (GE), on its transformation into a software company, has helped various clients in using big data's potential to their advantage for the varied set of benefits that it provides. For instance, one of their clients saved $7.5 million in losses due to unplanned downtime by simply replacing a piece of equipment before failure, due to the employment of big data [24]. Moreover, all the aforementioned real-world implementations are summarized in Table 10.5.

10.7   REAL-WORLD IMPLEMENTATION OF BIG DATA IN PETROLEUM REFINERY

Besides the research work, there has been real-world implementation of big data into refining operations as well. In a recent project by Repsol SA, big data analytics was used for conduction of management optimization for one of the company's integrated refineries in Tarragona (Spain). Google Cloud was used for providing Repsol with the data analytics products and consultation, while Google Cloud ML services would provide more insights on the same for this project. This project came with potential for economic gains as well as environmental issues' rectification [25].

Table 10.4 Real-world implementation of big data in reservoir studies

Companies	Implementation
SGS	Reservoir engineering services for O&G companies
Flotek	Transformation of reservoir management and recovery from reactive to predictive
Shell	Creation of oil reservoirs 3D and 4D maps to figure out the remaining content of oil and gas in reservoir

Table 10.5 Real-world implementation of big data in production and transportation

Companies	Implementation
Shell	Improvement of performance along with understanding of requirement of equipment maintenance
Shell	Understanding of refined products movement and effective fixing of product prices
Schlumberger	"Delfi"-based entire oilfield output maximization
General Electric	Big data–based software services to O&G companies

Table 10.6 Real-world implementation of big data in petroleum refinery

Companies	Implementation
Repsol	Management optimization for company's integrated refinery using big data analytics
Sinopec Yanshan Petrochemical Company	Real-time optimization for ethylene plant of the refinery using intelligent refinery 2.0
Shell	Real-time production optimization of major Martinez refinery installations
Shell	Global optimization through integration of operations in 17 refineries worldwide
Valero	Online utility optimization system–based efficiency improvements at Houston refinery

Real-time optimization of refinery-generated data has been on the rise in the industry in the last decade too. A petrochemical company named Sinopec Yanshan has implemented real-time optimization for its ethylene plant in the refinery, with an annual efficiency increase that has benefitted around 30–60 million CNY in profits. Shell has also implemented real-time optimizations in some of its major installations at Martinez refinery, providing a benefit of up to 10% profits (or cost save) per barrel in the 1990s [26].

Moreover, Shell has also utilized big data and intelligent refinery 2.0 (under intelligent oilfield 2.0) to integrate various operations (supply, treatment, transportation, sales and unified production) across its globally spread 17 refineries in order to achieve worldwide optimization, resulting in better emission prospects and economical gains [26]. Another company named Valero has approached efficiency improvements by the establishment of an online utility optimization system at its Houston refinery which led to a 0.6% increase in boiler thermal efficiency and 1% decrease in fuel gas costs [26].

Moreover, all the aforementioned real-world implementations are summarized in Table 10.6.

10.8 REAL-WORLD IMPLEMENTATION OF BIG DATA IN NATURAL GAS

Besides the research work, there has been real-world implementation of big data into petroleum natural gas operations as well. GE has used diagnostic analytics on its gas turbines (in Liquefied Natural Gas/LNG liquefaction and regasification plant).

Table 10.7 Real-world implementation of big data in natural gas

Companies	Implementation
General Electric	Big data–based diagnostic analytics on gas turbines
ExxonMobil	Acceleration of well planning and real-time AI insights-based optimization of drilling

It identified major factors to improve the efficiency of their plant as uptime, downtime and maintenance, where big data helps to increase profitability gas turbine to maximize the uptime and minimize the downtime. Moreover, as maintenance plays a major role in profitability of any industry, big data helps in avoidance of maintenance shutdowns (that might result in unexpected maintenance in equipment which could stop the production flow and disrupt the plant cycle) by using data analysis for prediction of maintenance and its time cycle [27].

As a vital (and demanding) aspect of natural gas happens to be "well planning", which satisfies reservoir engineer's requirements for production, ExxonMobil accelerates the well planning process (along with drilling optimization for the same) with big data–based real-time AI insights [20,28]. Their process happens to be the most cost-effective ones that satisfy real investment life-cycle(s) in a manner that are compatible with environmental, safety and regulatory needs as well.

Almost all the aspects of drilling, reservoir and production mentioned in the Sections 10.4, 10.5 and 10.6 respectively also apply to big data's real-world implementation in natural gas. Moreover, all the aforementioned real-world implementations are summarized in Table 10.7.

10.9 REAL-WORLD IMPLEMENTATION OF BIG DATA IN HEALTH AND SAFETY

Besides the research work, there has been real-world implementation of big data into petroleum health and safety operations as well. Several companies have put their efforts into the use of better and greener operations that reduce the risk of accidents while making the actual work environment safe. For instance, Shell aims to reach zero-emission energy system in their operations with the help of carbon capture and storage technology empowered by big data software [29].

Chevron, another O&G company, has strong claims on the estimated effectiveness of a fully optimized digital oil field for effective and efficient working environment, mentioning that it can also result in economic gains such as 8% higher production rates and 6% higher overall recovery rates. In lieu of this, Chevron's Tengiz oilfield in Kazakhstan has been made to include about 1 million sensors for the aforementioned benefits that it provides [23].

As discussed earlier, big data analytics has also been used for the conduction of management optimization for one of Repsol SA integrated refineries in Tarragona (Spain) with the help of Google Cloud Platform-aided ML services. The project did come with the potential for economic gains; however, it also came with environmental issues' rectification (that would ensure health and safety of operations' affected individuals) [25].

Table 10.8 Real-world implementation of big data in health and safety

Companies	Implementation
Shell	Long-term aim for zero-emission energy system
Chevron	Usage of fully optimized digital oil field on their sites
Repsol	Environmental rectification of company's integrated refinery using big data analytics
Woodside	Reduction in time taken for safety incident information analysis
Intertek	Big data research project for safer working environment
Bakken [All]	Drilling rig efficiency improvements incorporated through safe-practices

Moreover, IBM's Watson big data systems have ingested more than 30 years of Woodside's health and safety data with a reported 80% reduction in time taken for analysing safety incident information. The health and safety team used to receive nearly tens of thousands of safety alert cards a month at one of their gas plants, for which manual checks would take hours (or possibly days). For the same, big data solutions allowed them to do the analysis for obtaining feedback on the permit process, followed by provision of said insights to site controllers for appropriate actions and improvements [30].

There have been company-university projects for the exploitation of big data's potential into petroleum industry's operations as well. For instance, a company named Intertek collaborated with Robert Gordon (RG) University in the UK on a research project [31] that focused on use of big data for aiding oil and gas companies to improve asset performance, increase efficiency, reduce operating costs and develop safer working environment for workers.

Moreover, there has also been usage of big data for incorporation of safety measures into petroleum industry's operations. For instance, in another study by Duffy [32], the drilling rig efficiency was drastically improved by incorporating "safe-practices" identified using an "automated drilling state detection monitoring service". In their case study on Bakken pad drilling, their results provided savings of around 12 days along with the total non-drilling time reduction by observed 45% (almost reduction by half). Moreover, all the aforementioned real-world implementations are summarized in Table 10.8.

10.10   REAL-WORLD IMPLEMENTATION OF BIG DATA IN FINANCE AND FINANCIAL MARKETS

Besides the research work, there has been real-world implementation of big data into petroleum finance and financial markets as well. For instance, Bain & Company provides big data–based financial services to their clients as it was revealed that companies with better analytics capabilities were twice as likely to be in the top quartile of financial performance in their industry. In addition, as many companies prioritize financial tracking and reporting, barriers may exist between these and the operational systems that collect and manage important geologic and performance data. In other words, most companies design these systems for the financial needs

Table 10.9 Real-world implementation of big data in finance and financial markets

Companies	Implementation
Bain & Company	Financial services such as financial tracking and reporting among a set of other big data services
Shell	Big data–based financial aid to industry operations
Repsol	Big data–based financial risk minimization
Hitachi	Financial services for oil and gas industry companies
mapR Technologies	Financial services for oil and gas industry companies
Amazon, Google, Microsoft and Oracle	Time series prediction as a service under their cloud infrastructure

of the centre, rather than the operational needs of the field, and big data can be a tremendous help for the same [33].

Shell also uses big data for aid in a variety of their financial tasks such as industry-wide (exploration, production and transportation) cost or expenditure reduction as well as financial management of the intricate aspects related to them. Big data analytics helps in usage of complex algorithms to analyse economic factors throughout the industry, helping in other financial correlated tasks (product distribution and price setting) as well [22].

The Repsol SA's implementation of big data analytics for management optimization of one of its integrated refineries in Spain was also accompanied by pricing optimization and financial risk minimization with the help of big data analytics on the humongous amount to refinery and time series data in consideration, resulting in better economic upside [25].

There are various other big data-based companies that operate under the oil and gas industry while catering to the big data-based solution needs to O&G companies. Hitachi and mapR Technologies are two such companies that cater financial services to petroleum companies where their services range from financial modelling, asset management and even time-series financial markets prediction [34].

Several companies have started providing services for time series prediction as well (for clients if not for themselves). For instance, Amazon rolled out its own time series prediction service as recently as 2020 under forecasting as a service proposition while being just a single step in their data pipeline [35]. The case is similar for other companies like Google, Microsoft and Oracle which provide similar services under their cloud infrastructure. Moreover, all the aforementioned real-world implementations are summarized in Table 10.9.

10.11 CONCLUSION

This chapter establishes the fact that the implementation of big data in the petroleum industry is not just theory and research but actual practical component of big data's role in the entire industry. There has been significant development in the extent of real-world application of big data by O&G companies in the last few decades. By doing so, they have reaped the benefits that the implementation brings to the table, and the effects of the same are already visible through massive corresponding economic upsides.

Once it is known that big data is already in place in real world by various O&G companies, it is also vital to understand which aspects or traits of few companies make their implementation superior to other. This is what will be discussed in the upcoming chapter on superior traits of big data implementation which will help in the proper comprehension on what makes one implementation better than the other and how such aspects can be undertaken by almost all other O&G companies.

REFERENCES

1. Marr, B. (2015). Big Data In Big Oil: How Shell Uses Analytics To Drive Business Success. https://www.forbes.com/sites/bernardmarr/2015/05/26/big-data-in-big-oil-how-shell-uses-analytics-to-drive-business-success/.
2. Lucero, T. H. (2020). Three Innovative Ways Shell Is Using Big Data. https://businesschief.com/technology-and-ai/three-innovative-ways-shell-using-big-data-1.
3. Natkar, S. (2016). Big Value for Big Data in Oil and Gas Industry! https://www.esds.co.in/blog/big-value-big-data-oil-gas-industry/.
4. Cowles, D. (2015). Oil, Gas, and Data: High-Performance Data Tools in the Production of Industrial Power. https://www.oreilly.com/content/oil-gas-data/.
5. Lux Research (2015). As Oil Prices Plunge, Big Oil Turns to Big Data. https://www.luxresearchinc.com/press-releases/as-oil-prices-plunge-big-oil-turns-to-big-data.
6. Hassani, H., & Silva, E. S. (2018). Big data: A big opportunity for the petroleum and petrochemical industry. *OPEC Energy Review*, 42(1), 74–89. https://doi.org/10.1111/opec.12118.
7. Kambouris, T. (2016). How Big and Fast Data Can Transform the Oil and Gas Industry. https://www.datanami.com/2016/04/04/big-fast-data-can-transform-oil-and-gas-industry/.
8. Boman, K. (2015). Big Data, Internet of Things Transforming Oil and Gas Operations. https://www.rigzone.com/news/oil_gas/a/140631/Big_Data_Internet_of_Things_Transforming_Oil_and_Gas_Operations.
9. Enhance Reservoir Engineering with Simulation Tools. Infosys (2021). https://www.infosys.com/industries/oil-and-gas/industry-offerings/reservoir-engineering.html.
10. Baumann, P., Mazzetti, P., Ungar, J., Barbera, R., Barboni, D., Beccati, A., Bigagli, L., Boldrini, E., Bruno, R., Calanducci, A., Campalani, P., Clements, O., Dumitru, A., Grant, M., Herzig, P., Kakaletris, G., Laxton, J., Koltsida, P., Lipskoch, K., . . . Wagner, S. (2015). Big data analytics for earth sciences: The EarthServer approach. *International Journal of Digital Earth*, 9(1), 3–29. https://doi.org/10.1080/17538947.2014.1003106.
11. India WRIS (2019). https://indiawris.gov.in/wris/
12. Madhukar, M., & Pooja, A. (2019). Earth Science [Big] Data Analytics. https://doi.org/10.1007/978-3-319-89923-7_4.
13. Climate Home. NASA (2008). https://climate.nasa.gov/.
14. Li, Z., Yang, C., Jin, B., Yu, M., Liu, K., Sun, M., & Zhan, M. (2015). Enabling big geoscience data analytics with a cloud-based, MapReduce-enabled and service-oriented workflow framework. *PLoS One*, 10(3), e0116781. https://doi.org/10.1371/journal.pone.0116781.
15. Li, Z., Yang, C., Sun, M., Li, J., Xu, C., Huang, Q., & Liu, K. (2013). A High Performance Web-Based System for Analyzing and Visualizing Spatiotemporal Data for Climate Studies. *Web and Wireless Geographical Information Systems*, 190–198. https://doi.org/10.1007/978-3-642-37087-8_14.
16. Yang, C., Li, W., Xie, J., & Zhou, B. (2008). Distributed geospatial information processing: Sharing distributed geospatial resources to support digital earth. *International Journal of Digital Earth*, 1(3), 259–278. https://doi.org/10.1080/17538940802037954.
17. Smith, K. (2016). The Impact of Big Data, Open Source on Oil and Gas. https://www.hart-energy.com/exclusives/impact-big-data-open-source-oil-and-gas-29465.

18. Bolen, M., Crkvenjakov, V., & Converset, J. (2018). The Role of Big Data in Operational Excellence and Real Time Fleet Performance Management – The Key to Deepwater Thriving in a Low-Cost Oil Environment. *IADC/SPE Drilling Conference and Exhibition 2018*, 1–19. https://doi.org/10.2118/189603-ms.

19. Reservoir Engineering Services. SGS Group (2018). https://www.sgsgroup.in/en-gb/oil-gas/upstream/subsurface-consultancy/specialized-studies/reservoir-engineering-services.

20. Use Analytics and AI to Turn Big Data for Oil and Gas into Insights. Big Data for Oil and Gas Case Studies, IBM (2020). https://www.ibm.com/industries/oil-gas/big-data-analytics.

21. Flotek Presentation: Improving Production through Reservoir-Centric Fluid Systems. Flotek (2019). https://www.flotekind.com/index.php/newsroom/item/1176-flotek-presentation-improving-production-through-reservoir-centric-fluid-systems.

22. Bekker, A. (2020). How to Benefit from Big Data Analytics in the Oil and Gas Industry? https://www.scnsoft.com/blog/big-data-analytics-oil-gas.

23. Crooks, E. (2018). Drillers Turn to Big Data in the Hunt for More, Cheaper Oil. https://www.ft.com/content/19234982-0cbb-11e8-8eb7-42f857ea9f09.

24. ARC Advisory Group (2014). GE Minds and Machines 2014 – The Customer Perspective. https://www.arcweb.com/ja/node/12221.

25. Mohammadpoor, M., & Torabi, F. (2020). Big data analytics in oil and gas industry: An emerging trend. *Petroleum*, 6(4), 321–328. https://doi.org/10.1016/j.petlm.2018.11.001.

26. Lu, H., Guo, L., Azimi, M., & Huang, K. (2019). Oil and gas 4.0 era: A systematic review and outlook. *Computers in Industry*, 111, 68–90. https://doi.org/10.1016/j.compind.2019.06.007.

27. Patel, H., Prajapati, D., Mahida, D., & Shah, M. (2020). Transforming petroleum downstream sector through big data: A holistic review. *Journal of Petroleum Exploration and Production Technology*, 10(6), 2601–2611. https://doi.org/10.1007/s13202-020-00889-2.

28. Newman, H. E., Roberts, C. M., & Biegler, M. W. (2005). How Integrated Well Planning, Technology, and Operations Excellence Impact ExxonMobil's Development Well Results. *SPE/IADC Drilling Conference 2005*, 1–6. https://doi.org/10.2118/92198-ms.

29. Shell (2018). Net Carbon Footprint. Shell Sustainability Report 2018. https://reports.shell.com/sustainability-report/2018/sustainable-energy-future/net-carbon-footprint.html.

30. Woodside. Data Science (2015). https://www.woodside.com.au/what-we-do/innovation/data-science.

31. RG University (2016). Intertek and RGU Help Oil and Gas Companies Make the Most of Big Data. https://www.agcc.co.uk/news-article/intertek-and-rgu-help-oil-and-gas-companies-make-the-most-of-big-data.

32. Duffy, W., Rigg, J., & Maidla, E. (2017). Efficiency Improvement in the Bakken Realized through Drilling Data Processing Automation and the Recognition and Standardization of Best Safe Practices. *SPE/IADC Drilling Conference and Exhibition 2017*, 1–10. https://doi.org/10.2118/184724-ms.

33. Padmanabhan, V (2014). Big Data Analytics in Oil and Gas. Bain & Company. https://www.bain.com/contentassets/38df79e1ec42486497d197c48e09118b/bain_brief_big_data_in_oil_and_gas.pdf.

34. Technavio (2014). Top 8 Big Data Companies for the Oil and Gas Sector. *Technavio Blog*. https://blog.technavio.com/blog/top-8-big-data-companies-for-the-oil-and-gas-sector.

35. O'Reilly Media (2020). The Future of Time Series Forecasting. Medium. O'Reilly. https://medium.com/oreillymedia/the-future-of-time-series-forecasting-bd83c2aca9a8.

Chapter 11

Traits of Companies with Superior Big Data Implementation

11.1 INTRODUCTION

There is a massive amount of research work pertaining to big data's implementation in the petroleum industry, and its effectiveness can be conferred by the corresponding number of real-world applications of big data by O&G companies. Keeping this in mind, there is an essential topic that needs to be discussed – what makes few companies better at big data implementation than others?

There are indeed some aspects of the selected few that make their implementation "superior", which is exactly what is discussed in this chapter. In a way, the chapter can also be considered as a cumulation of best practices that a company should undertake or ensure in order to make the best out of big data implementation.

Few traits of successful big data implementation include clarity of end goal, clear employee outlook and resource management. A deeper dive into all of them will be done in the next section, which elaborates all the facets of superior implementation (Section 11.2), followed by a conclusion (Section 11.3).

11.2 MAJOR TRAITS OF COMPANIES WITH SUPERIOR BIG DATA IMPLEMENTATION

Despite the positive outlook of the implementation of big data in the industry, a "successful" implementation of big data is often not achieved due to various traits pertaining to the company implementing it. Therefore, the hereby mentioned traits need to be on the focus of every company that plans integration of big data into any petroleum operations.

The first and foremost trait that every company must uphold is the close and mutual collaboration between the team (that operates or handles big data integration) and the business owners (who aim to utilize big data's potential but do not fully understand the intricacies of it). This ensures a smooth, hassle-free, crystal-clear and effective working (and thus implementation) of big data and its analysis into various operations.

There should (if not must) be a non-hierarchical structure as well as culture for everyone involved in big data integration (data scientists, team workers, researchers and business owners), which ensures that the people who understand and are directly involved in implementation feel empowered for their roles and take complete ownership of their work.

DOI: 10.1201/9781003185710-11

Furthermore, a company that understands the potential of big data should also be willing to invest time and resources in the innovations required for their specific scenario as well as foundational technology for the same. The company must not rely on short-term ROI (return on investment) and proactively invest in expertise, tools and technologies that pose long-term benefits.

Along with the advance into newer technology, there should be clear decision rights across older legacy investment technology and newer investment technologies. The company must be willing to make sacrifices when the older investments do not pose long-term benefits while the newer ones do. In such case, the company must be prepared to "let go" of the previous technological investment(s) and envision the future.

Moreover, a simple yet vital trait about a company which plans on implementing big data is the clarity of "what's the end goal?" A major mistake that several companies make is the upfront investment and initialization of implementation, without proper set of pre-defined goals for the same. This should not happen in any case for successful integration of big data, as having a stark clear understanding of where big data needs to be applied and in which order (based on operational and business needs) along with a proper set of milestones for measuring success benefits companies enormously.

In order to make the engineering and integration seamless without any human-based social hurdles, the roles and responsibilities of each and everyone associated with implementation should be clearly communicated to avoid any chance of the aforementioned issues from the very beginning. Also, the need for new roles or updates (in roles and responsibilities) should also be communicated properly and considered equally important.

Moreover, as there is a significant lack of professional expertise in the industry, there must be clear career pathways for each role associated with big data, and these pathways need to be openly and clearly communicated to every individual on board. Doing so ensures that there is a proper retention of talent in the team, and they can focus on their work efficiently and effectively without worrying about their future career.

The company must also ensure that their big data team follows engineering and code deployment standards (or principles), which helps in effective test and releases. There must be continuous delivery of all aspects related to work to reduce rework, decrease project time and increase speed & efficiency. Major engineering work-related issues are tackled here if this trait is present and undertaken successfully.

Another interesting and overlooked aspect also needs to be upheld, which is proper utilization of automation tools while working on tasks that are repetitive and can be automated. Ensuring that the experts do not waste their time on such tasks ensures that they spend time on quality work (such as actual data analysis and research). Moreover, it also helps in drastically ramping up the speed of various processes and tasks such as code-base testing, model deployment, data cleaning (majority of it) and result (analysis outcome) publishing.

A proper resource management system should also be in place as a major company trait as it makes sure that proper resources are allocated across expertise, long-term and annual-term bases. Doing so reduces the chances of resource scarcity in future while ensuring that everyone on board has clear understanding of available resources (and thus work accordingly).

Finally, once a well-thought resource management is in place, an appropriate centralization for the allocation of these resources (and performing necessary management

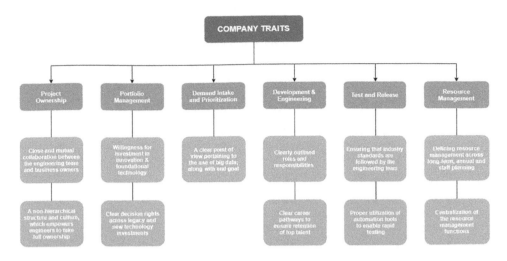

Figure 11.1 Traits of companies with successful big data implementation.

functions) needs to be taken care of. Doing so will ensure that despite proper resource management plan, it is upheld properly too. All the aforementioned traits are condensed in the Figure 11.1 for better and quick comprehension.

11.3 CONCLUSION

A cumulation of all the traits (mentioned in the previous section) in a company that wants to implement big data in its operations has tremendously higher chances of successful integration due to the benefits mentioned along with the individual traits compared to its counterpart(s). Most of these traits (if not all) are exemplified by the actual O&G companies which portray them.

These traits also bring forth various benefits along with them that make undertaking them worthy, including:

- Better asset and resource management
- Improved efficiency and effectiveness
- Collaborative and lively work culture
- Upholding industry standards

However, despite the advantages and the said traits for superior implementation, there are various hurdles along the way that restrict exploitation of complete potential of big data in the industry, which happens to be the topic of discussion in the next chapter on challenges of implementation of big data. Exploiting the best of the capabilities of big data by overcoming the said challenges, O&G companies can see drastic improvements which would else seem impossible to achieve even with the current big data standards.

Chapter 12

Challenges of Big Data

12.1 INTRODUCTION

Throughout the course of the book, the chapters have portrayed a quite optimistic outlook to big data in the industry. The benefits mentioned are massive and possess capability to revamp the entire industry, and the same goes for the diverse range of applications that help O&G companies in un-fathomable circumstances and areas of execution. The research work and real-world implementation until now also look promising. Thus, the question that arises is – why is there not full-fledged big data implementation in the industry?

The simple answer to that is the existence of problems, hurdles or challenges along the way that restrict or block the actual progress in big data's use in the industry by companies, and they come in all levels and types. Hence, it is important to comprehend what the challenges are, what issues they cause and how (and why) they exist in the first place.

Few challenges of real-world big data implementation include lack of expertise (and investment in the same), decision-based hurdles and economic & resource–related limitations. A deeper dive into all of them will be done in the next section, which elaborates all the facets of major industry-wide challenges (Section 12.2), followed by a conclusion (Section 12.3).

12.2 MAJOR CHALLENGES

One major hurdle that plagues every company is the difficulty in transferring of data generated from thousands of sensors to data servers or processing facilities (as a part of various workflow operations) that perform analysis upon it to provide meaningful insights as output. The segregation of data based on data type, data amount and data protocols (mostly undertaken on premises) also poses another essential challenge for the effective and seamless functioning of petroleum operations before the transfer-related challenge(s) ever come up.

Another major hurdle happens to be the decision-making process on defining the frequency of collecting the data from the sensors in order to gain optimum result (real-time n-dimensional geological maps). This includes understanding of how different frequencies of data collection affect the performance while keeping the frequency as low as possible (to increase sensor's lifespan and reduce data transmission bandwidth). Thus, a rigorous process of balancing the trade-offs between

DOI: 10.1201/9781003185710- 12

frequency and performance is generally done before starting and while working on final long-term operation(s). Besides, the same process is also undertaken for transferring of data from on-premise storage facility to cloud-based server facilities as well (majorly in order to reduce cloud service costs).

There are also issues with latency (delay in transmission from sender to receiver) of data moving back-and-forth from sensors to on-premise storage facilities to analysis facilities (that hinder "real-time" output). Although most of the applications for big data do not rely on real-time data input, few operations (such as reservoir engineering, natural gas treatment and extraction) may have detrimental effects in its performance and functioning once the latency reaches a certain threshold. Although the use of cloud services (and cloud computing), fog computing (ability to compute, make decisions and take action on IoT devices, and only push relevant data to the cloud) and faster IoT devices reduce the latency to a certain extent, there has not been a perfect solution yet.

Furthermore, besides the frequency of data, the quality of data also matters in optimizing the insights gained from big data analytics. And thus, another challenge turns out to be in the decision-making process of sensor purchase while defining the quality of sensory data transmitted to on-premise facilities (and processed quality of data sent to the cloud servers) from on-premise (production, drilling, etc.) and off-premise (transportation, refinery, etc.) locations. The balancing of trade-offs happens to be more-or-less the same as in the frequency-based hurdle and happens to portray a major economic and performance-based challenge.

Along with the quality of data, the consideration or use of proper analysis tool and model(s) is also a major challenge, as despite quality data, an analysis model can make or break an analysis insight or report. There are hundreds (if not thousands) of analysis models and tools out there, and choosing the best one can be a time-consuming and tedious task and thus require expert professionals on the team to make effective decisions on model usage through knowledge and experience.

Furthermore, even when the challenges pertaining to data quality and analysis models are taken care of, another vital economic hurdle comes into picture. With such amount of data comes an incredible need of computational resources which can be detrimental to expenditure limits if legacy infrastructure is used. Despite the emergence of cheap computational devices as well as highly accessible cloud service, the sheer amount of data handling and its required processing power can cause major setbacks to companies' investment budget.

There are governmental regulatory limits to the execution (of geoscience, production, transportation, drilling and refining activities) that differ in each country. These regulatory limitations are majorly set in place due to environmental concerns and to ensure that geoscientists do not go beyond a certain threshold in their carbon footprint and greenhouse emissions. This affects the associated operations in terms of:

- transportation routes and means
- limitations in involved equipment
- working hours
- processes used

which prove to be a major hurdle to solve while keeping greener solutions in mind.

Moving forward, there are scientific challenges to big data's implementation as well. For instance, almost all of the operations of oil and gas require thorough understanding of physics, mathematics and statistics. Therefore, for obtaining meaningful and optimum results from big data, the petroleum engineers should collaborate or involve data scientists in order to make sure that right big data tools and procedures are utilized for analysis. Doing so helps in ensuring correct and optimized analysis insights & solutions from the same data, which eventually saves tremendous amount of time (on previous hit-and-trial activities) and money.

Having mentioned that, there is also a rising need for engineers (such as reservoir engineers in reservoir studies) that specialize in cloud technologies, edge computing, software development and data science (professionals equipped with technical know-hows in the big data field) along with specialization in specific operations. This is in accordance with need for expertise required for big data integration in various aspects of operations that would turn out to be pretty expensive if contractors are hired for the same. Therefore, companies have started to employ professionals who have expertise in both petroleum and big data industry to offset economic costs. For instance, Shell employs 70 full-time working individuals in their data-analysis department along with hundreds of other experts across the globe on an ad-hoc basis. However, despite such hiring decisions, the supply of trained professionals is always less than what is required in the industry at large, as mentioned in the first chapter as well, thus posing a major challenge.

On a side note, the expertise required by a data scientist for working effectively in oil and gas industry is also extensive, and include the following tools and technologies:

- Hadoop ecosystem along with Apache Spark (for large data storing and processing).
- Snowflake, AWS EMR, AWS Ethena, GCP Cloud Composer, AWS Redshift, Hive, Azure Synapse Analytics, GCP BigQuery Dataprots, Azure DataFactory or DataBricks (cloud-based big data tools).
- SQL/NoSQL databases.
- MapReduce old Java code Maintenance.
- Scala and Python (newer programming languages).
- Kubernetes (building Big Data CI/CD pipelines).
- Kafka, AWS Kinesis or Apache Pulsar (for real-time big data streaming).

Seeking individuals with such extensive knowledge of tools and technologies is often time-consuming and an expensive affair. Hence, the "expertise-angle" of oil and gas industry turns out to be one of the major challenges. Moreover, due to the fact that low petroleum prices [1] have been deterring the O&G companies, they do not promote any form of investment in data science professionals who can employ big data for the companies' benefit [2]. Furthermore, this neglect in adopting big data has (and will have) negative impacts on petroleum industry's growth that will only compound over the years [3].

The cumulation of these hurdles portrays the reason why the oil and gas industry has been slow and uninterested in big data's implementation for decades. However, the tide is turning now as there's heightened need for efficiency improvements, environmental impact reduction, ensuring of safety and cost optimization due to various factors mentioned in the first chapter, resulting in acceptance of big data in recent years. However, the level of acceptance and integration remains to be seen in the upcoming decades. Moreover, all the challenges discussed in this section are summarized in Figure 12.1.

Figure 12.1 Major challenges of big data's implementation in petroleum operations.

12.3 CONCLUSION

The challenges faced by the O&G companies are the backbone of what restricts them to wholeheartedly accept and undertake big data implementation throughout their operations. There are various entries mentioned in the previous sections that are on the verge of being solved while others are far from being solved, and the solution to them provides massive benefits including:

- Better talent and expertise (and hence implementation)
- Smooth and hassle-free execution of operations
- Improved efficiency and effectiveness of operations
- Reduced expenditure and other economic costs

However, despite these hurdles and challenges that bother the entire industry, there are several positive aspects of future prospects that portray an optimistic picture on where the big data implementation can go. This is precisely the topic of discussion in the next chapter on future scope of big data. Taking the said future prospects into consideration, O&G companies can plan on to reap the benefits of the same in the long term with proper planning, structuring and focus on big data's role in their operations.

REFERENCES

1. Dodgson, L. (2016). Cutting Costs: How Big Data Can Help. https://www.offshore-technology.com/features/featurecutting-costs-how-big-data-can-help-4733715/.
2. Kivi, J. (2017). How the Internet of Things has Influenced Midstream Pipeline Operations. https://blog.schneider-electric.com/industrial-software/2017/10/03/iiot-midstream-pipeline-operations/.
3. Dyson, R. (2016). The Oil and Gas Industry Needs to Think Big with Big Data. https://www.worldoil.com/sponsored-content/digital-transformation/2016/17/the-oil-and-gas-industry-needs-to-think-big-with-big-data.

Chapter 13

Future Scope of Big Data

13.1   INTRODUCTION

The previous chapter brought forward the existence of problems, hurdles or challenges that block the actual progress in big data's use in the industry by companies, after a complete course of optimistic outlook in the book until then. Now that they are understood properly, it is time to dive into the future prospects of big data that pose as massive upsides to its real-world implementation, and if O&G companies undertake them effectively, they can reap the benefits that it offers in the long term.

Few future prospects of big data's long-term future scope include feasibility of big data platform & tools' usage and increase in effectiveness & efficiency of operations. The deeper dive into all of them will be done in the next section, which elaborates all the scopes of big data in the future (Section 13.2) followed by a conclusion (Section 13.3).

13.2   MAJOR ASPECTS OF FUTURE SCOPE

Given the benefits, applications, implementation, research boom and challenges of big data in the industry throughout the previous chapters, there are various factors that pose towards an optimistic future of big data's integration into the industry. The hereby mentioned factors happen to be in the limelight while having major aspirational future prospects.

An interesting fact to understand about petroleum industry's big data is that it is not truly big data yet, because currently, only 3%–5% of petroleum industry's equipment is connected [1] and hence the big data–based tasks (such as drilling, transportation and financial analysis) are not worthwhile. And considering another fact that just exploiting the true potential of current big data can provide 20% operational efficiency [2]; ramping up the 3%–5% connectivity can have tremendous positive upsides in the future.

The increase in cloud service's usage, viability and cost effectiveness has been on the rise in the last decade. Currently, the platforms like Amazon Web Services, Google Cloud Platform and Azure (by Microsoft) have invested heavily into making the use of cloud services as cheap, easy and accessible to everyone as possible. Moreover, the previously mentioned fact that complex and computationally extensive analyses can be conducted with ease in lesser time on such platforms is a major bonus. These advances make the adoption of cloud-based big data implementation (storage, maintenance and analysis) feasible for the entire industry.

DOI: 10.1201/9781003185710-13

Moreover, despite of the fact that technology has been used in petroleum industry for decades, it generally brought expensive installation costs along with it that made it uneconomical for usage in real-world scenario. This is however not the case now, as sensors have now become inexpensive and lightweight (coupled with safe and secure wireless networks) making them feasible to use along with big data analytics for gaining meaningful insights for better decision making in all petroleum operations. And this trend of increase feasibility and inexpensive equipment will only keep rising in the future.

Also, it is obvious to note and take the advances in big data itself over the year into consideration, and outline the possibilities of its growth in the upcoming decade. The rise and advances in tools like Hadoop, Apache Storm, MongoDB and Cassandra do not seem to slow down, and the improvements they consistently bring can effectively keep benefiting the industry with lower operational costs, less downside risks and less uncertainty. Moreover, there is an increase in real-world applicable research projects too, as depicted in the research chapter earlier. Therefore, there is an inherent advantage in the future scope of big data.

There is an increase in "real-world applicable research projects" as well. For instance, a company named Intertek collaborated with Robert Gordon (RG) University in the UK on a research project [3] that focused on the use of big data for aiding oil and gas companies to improve asset performance, increase efficiency, reduce operating costs and develop safer working environment. Another instance is the introduction of analysis system called Petroleum Analytics Learning Machine that is based on big data–employed machine learning and provides real-time analysis of thousands of Internet of Thing (IoT) sensors [4] at Institute of Electrical and Electronics Engineers (IEEE) International Conference on Big Data 2017. Such instances of real-world research projects outlay solid future likelihood for big data's implementation in the industry.

It was noted in a survey by International Data Corporation (IDC) Energy that the major hurdle for big data in the industry is the lack of awareness regarding the benefits of the topic besides its proper understanding in the industry, resulting in a miniscule business support [5]. This however is changing now as another study by Mehta [6] after 5 years (in 2018) shows that 81% of volunteering executives considered big data to be one of the top 3 priorities for their O&G company in the upcoming years. This clear rise in awareness and acknowledgement of big data is surely something to look out for in the future.

It should also be noted that despite of obvious big data benefits in petroleum industry like "model-based beforehand" efficiency improvements and real-time data analysis, other benefits like near-real-time visualizations, data storage (and management) and near-real-time anomaly detection (for improved safety) happen to be the most important upsides of using big data, which are otherwise not possible to achieve [7]. These aspects make big data's implementation future prospects vibrant and feasibly likely to happen.

Moreover, considering the fact that the petroleum industry is notorious for lagging behind other industries in terms of technology's adoption and integration [8], the scenario is the same in terms of big data's implementation to its full potential as well. This however presents an opportunity and future scope of its adoption in the upcoming decade due to various aforementioned benefits as well as heavy research dedicated for the same, as it has a big wide room to grow in almost all petroleum operations.

On a side note, although one of the challenges mentioned earlier state that the low petroleum prices [9] obstruct investment in data science personnel in the industry, few

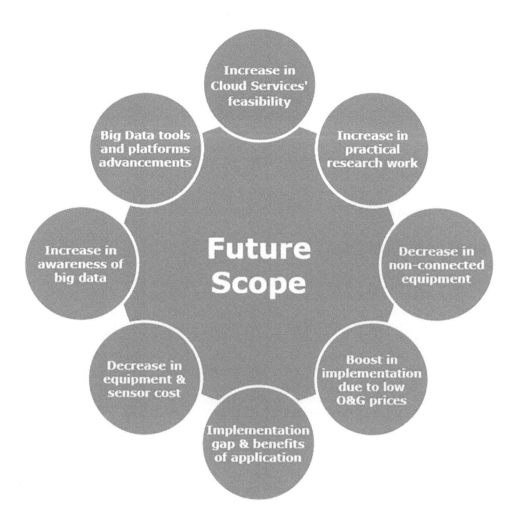

Figure 13.1 Future scope of big data's implementation in petroleum operations.

[1,2,10,11] also argue that the low prices might actually promote the investment into big data (and pose as a big future opportunity) so as to make better and constructive operational decisions. These are contrasting views onto a single situation, but this upside along with an amalgamation of all the aforementioned prospects portrays an extraordinary future potential like no other. Moreover, Figure 13.1 summarizes all the aspects pertaining to future scope of big data.

13.3   CONCLUSION

Despite the challenges faced by the O&G companies that are backbone of what restricts them to wholeheartedly accept and undertake big data implementation throughout their operations, this chapter has established the fact that there are indeed various future scopes that negate the non-positive aspects of challenges which result into long-term upsides to big data's real-world implementation. Moreover, the said future scopes provide massive benefits that include:

- Increase in feasibility of big data platforms' use
- Improved efficiency and effectiveness of operations
- Reduced expenditure and other economic costs
- Increase implementation and aforementioned benefits

Moving forward, it can be overwhelming to understand and comprehend various aspects of "big data in petroleum streams" when the information provided in the book is zoomed out. Thus, in order to provide a comprehensive summary to the book's contents, the topic of discussion for next (and last) chapter is the conclusion of the book which covers the gist of complete big data technology in the Oil and Gas industry.

REFERENCES

1. Mathew, B. (2016). How Big Data Is Reducing Costs and Improving Performance in the Upstream Industry. https://www.worldoil.com/news/2016/12/13/how-big-data-is-reducing-costs-and-improving-performance-in-the-upstream-industry.
2. Hassani, H., & Silva, E. S. (2018). Big Data: A big opportunity for the petroleum and petrochemical industry. *OPEC Energy Review*, 42(1), 74–89. https://doi.org/10.1111/opec.12118.
3. RG University (2016). Intertek and RGU Help Oil and Gas Companies Make the Most of Big Data. https://www.agcc.co.uk/news-article/intertek-and-rgu-help-oil-and-gas-companies-make-the-most-of-big-data.
4. Anderson, R. N. (2017). "Petroleum Analytics Learning Machine" for Optimizing the Internet of Things of Today's Digital Oil Field-to-Refinery Petroleum System. *2017 IEEE International Conference on Big Data (Big Data)*, 4542–4525. https://doi.org/10.1109/bigdata.2017.8258496.
5. Feblowitz, J. (2013). Analytics in Oil and Gas: The Big Deal about Big Data. *SPE Digital Energy Conference and Exhibition 2013*, 1–6. https://doi.org/10.2118/163717-ms.
6. Mehta, A. (2016). Tapping the value from big data analytics. *Journal of Petroleum Technology*, 68(12), 40–41. https://doi.org/10.2118/1216-0040-jpt.
7. Zaidi, D. (2018). Role of Data Analytics in the Oil Industry. https://towardsdatascience.com/here-is-how-big-data-is-changing-the-oil-industry-13c752e58a5a.
8. Endress, A. (2017). Big Data Analytics Burst onto Upstream Industry as Next Frontier. https://www.worldoil.com/blog/2017/04/17/big-data-analytics-burst-onto-upstream-industry-as-next-frontier.
9. Dodgson, L. (2016). Cutting Costs: How Big Data Can Help. https://www.offshore-technology.com/features/featurecutting-costs-how-big-data-can-help-4733715/.
10. Boman, K. (2015). Big Data, Internet of Things Transforming Oil and Gas Operations. https://www.rigzone.com/news/oil_gas/a/140631/Big_Data_Internet_of_Things_Transforming_Oil_and_Gas_Operations.
11. Dhunay, N. (2016). Big Data's Next Big Impact? *Oil & Gas*. https://dataconomy.com/2016/04/big-datas-next-big-impact-oil-gas/.

Chapter 14

Conclusion

14.1 INTRODUCTION

This chapter can be considered as a summarized version of the previous chapters that almost provides comprehension of all aspects of big data in the industry including 6Vs, benefits, applications, implementation, research, AI algorithms, big data platforms, real-world implementation, superior implementation traits, challenges and future scope.

A glance into all of them will begin with the next section starting with petroleum operations (Section 14.2), 6Vs (Section 14.3), benefits (Section 14.4), applications (Section 14.5), implementation (Section 14.6), big data platforms (Section 14.7), AI algorithms (Section 14.8), research (Section 14.9), real-world implementation (Section 14.10), superior implementation traits (Section 14.11), challenges (Section 14.12) and future scope (Section 14.13) followed by Section 14.14 marking end of the book.

14.2 PETROLEUM OPERATIONS

Geoscience, also known as Earth Science, is the study of Earth and includes so much more than rocks and volcanoes, as it comprises processes that form and shape Earth's surface, the natural resources we use and how water and ecosystems are interconnected with each other. It incorporates several scientific disciplines related by their applications to the study of the earth. In essence, it encompasses study and work with minerals, soils, energy resources, fossils, oceans and freshwater, the atmosphere, weather, environmental chemistry and biology and natural hazards among several others [1,2].

Exploration and drilling are one of the major operations in the upstream segment of petroleum production in oil and gas industry, and mark a vital point in the entire workflow of industry's processes. Exploration refers to the activities (research, analyses, survey and seismology) done for the search of viable source of hydrocarbons; and is deemed a high-risk operation that comes with massive expenditure. Drilling, on the other hand, is the operation that begins once a feasible source is located. As the name suggests, it refers to the actual drilling (or development) of sites (off-shore or on-shore) for the extraction of hydrocarbons underneath the surface.

Reservoir studies, also known as reservoir engineering, refers to a branch of study/ working under petroleum industry that deals with the scientific principles and applications of hydrocarbon flow through porous subsurface medium during the drilling

DOI: 10.1201/9781003185710-14

(development) and production operations from petroleum reservoirs in order to obtain a high economic recovery. The study consists of (but not limited to) subsurface geology, applied mathematics, technological integration processes and basic laws of physics & chemistry.

After the petroleum exploration has taken place, drilling process has been completed and reservoir modelling is successful, the next essential steps happen to be the actual production of hydrocarbons from the reservoir and its transportation via midstream operations to downstream operations' locations. The production of hydrocarbons is no easy feat (in terms of expertise, complexity of operation and level of execution-precision required) and involves several hazards such as reservoir leakage that make it risky affair. The same is true with transportation too, as it requires special expertise, solving of complex problems (such as cost minimization and route fixation) and is surrounded by hazards (especially petroleum leaks). This concerns everyone involved as the main goal here is to reduce the complexity of operations while minimizing the risks associated with it as much as possible.

After the production and transportation operations have taken place, the next essential step happens to be the refining of the transported hydrocarbons from the reservoir. The refining of hydrocarbons is no easy feat (in terms of expertise, complexity of operations and level of execution-precision required) and involves several activities (such as operation optimization and cost minimization) and hazards (such as refinery leakage) that make it a complex and risky affair [3]. This concerns everyone involved as the main goal here is to reduce the complexity of operations while minimizing the risks associated with it as much as possible (while taking environmental aspects into consideration).

Other than the crude oil gained from petroleum, a major product for petroleum is the gas refined from the raw hydrocarbons during the refining operations; which comprises of methane, natural gas liquids (which are also hydrocarbon gas liquids) and non-hydrocarbon gases (such as carbon dioxide and water vapour). It has an absence of colour, odour & taste and is non-toxic in nature; and its generation of natural gas occurs below the earth surface under extreme conditions in a petroleum reservoir.

During the execution of all the operations in oil and gas industry, an essential step happens to be the health and safety of the petroleum professionals who interact with the industry in one form or the other. The activities pertaining to ensuring petroleum health and safety is no easy feat (in terms of expertise, complexity of involved operations and level of execution-precision required) and involves several aspects (mentioned later) and hazards (also discussed later) that make it a complex yet life-saving affair [3]. This concerns everyone involved as one of the main goals of petroleum industry is to ensure minimization of the operation associated risks and safety improvements as much as possible (while taking environmental aspects into consideration).

Being one of the biggest sectors or portions of all industries in the world in terms of monetary value (as high as $2 trillion per annum in 2021 [4]), petroleum industry provides the most crucial economic framework – Oil to the world. Considering the fact that the industry today quenches 57% of global energy consumption, while roughly accounting for the same percentage of global CO_2 emissions, it requires intense capital, expensive equipment and highly skilled labour [5], all of which require extensive economic assistance, provided by petroleum finance and financial markets! Moreover, all of the operations are summarized in Figure 14.1.

GeoScience   Exploration   Drilling   Reservoir Studies   Production

Transportation   Refining   Natural Gas   Health & Safety   Finance & Markets

Figure 14.1 Petroleum operations.

14.3 6VS OF BIG DATA

Big data, in simple terms, can be attributed to a set of processes such data acquisition, data storage, search, analysis and visualization, whose cumulation can result in deep understanding of data at hand in order to make meaningful and educated decisions in the real world. To understand it intricately, one needs to understand different Vs associated with it, i.e., volume, variety, velocity, value, veracity and variability [6,7], which are summarized in Figure 14.2.

14.4 BENEFITS

After understanding the 6Vs of big data, it should be known that its usage brings a cumulation of results like deep understanding of data at hand in order to make meaningful and educated decisions in the real world, which pose massive benefits on the application field that are the reason for the boom in big data implementation (summarized in Figure 14.3).

14.5 APPLICATIONS

After diving into different aspects (6Vs) of big data in petroleum industry's each operation along with deeper dive into the benefits that implementation of big data can bring, understanding the scope of application of big data in the industry is also integral in understanding the role of big data in the industry. The major applications for the same are summarized in Figure 14.4.

14.6 IMPLEMENTATION

As inferred from the massive benefits and wide array of applications of big data, it certainly helps in deep understanding of data at hand in order to make meaningful and educated decisions in the real world. This simple aspect has led to widespread

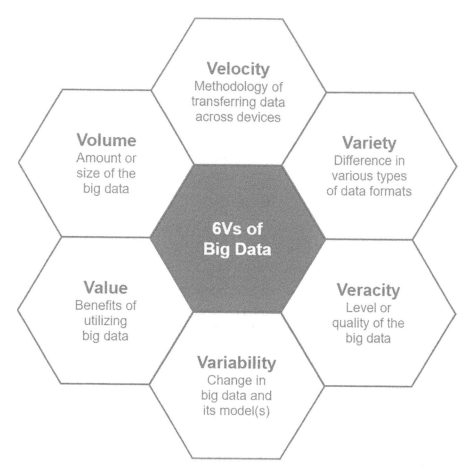

Figure 14.2 The 6Vs of big data.

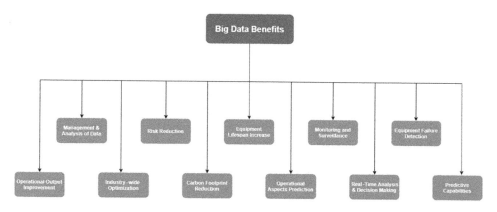

Figure 14.3 Industry wide benefits of big data.

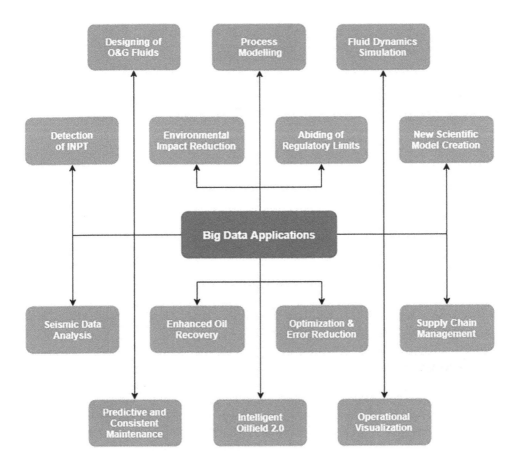

Figure 14.4 Industry wide applications of big data.

implementation of big data in every industry including oil and gas, whose standardized workflow is summarized in Figure 14.5.

14.7   BIG DATA PLATFORMS

During execution of big data–based petroleum operations, once the data has been transferred to cloud services, it needs to be analysed as well, mostly using big data platforms. There are various platforms in the marketplace, but the following turn out to be the major and widely used ones:

- Hadoop Infrastructure
- IBMPureData
- IBM InfoShpere
- Microsoft MURA
- Oracle Architecture Development Process (OADP)

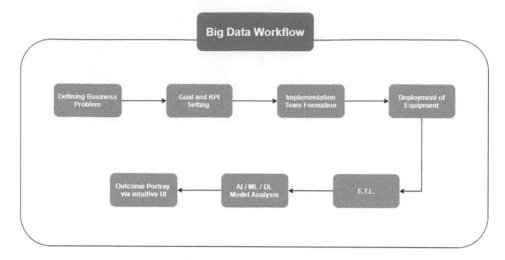

Figure 14.5 Big data workflow in petroleum operations.

Moreover, the usage of these big data platforms brings several benefits along with them, which are summarized in Figure 14.6.

14.8   AI ALGORITHMS

Although there are hundreds (if not thousands) of AI algorithms out there, which help in specific prediction and analysis scenarios (like classification, regression, decision trees, reinforcement learning, etc.), some are used more than others, and Figure 14.7 summarizes the major algorithms used in petroleum operations.

14.9   RESEARCH

A salient aspect of any field happens to be the amount of research work that is put into it, and considering the advent of big data along with the benefits it imparts on diverse applications, it is rather obvious to note that there has been significant contribution from the research community on the research work upon big data in various associated operations, which are summarized in Table 14.1.

14.10   REAL-WORLD IMPLEMENTATION

The success of significant contribution of the research community on advancements in big data research in each petroleum industry operation can be validated by assessing the level of real-world implementation undertaken by the O&G companies in the real world. It aids in comprehending whether the benefits along with technological advancements brought forward by the research are actually worth in the economic and conservative world of oil and gas. Therefore, Table 14.2 enumerates the major real-world applications of big data into petroleum industry by O&G companies.

Figure 14.6 Benefits of big data platforms.

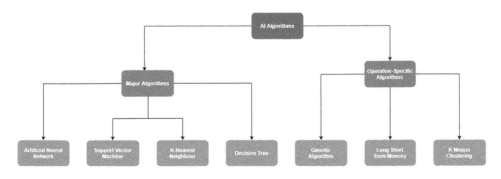

Figure 14.7 Major AI algorithms with operation specific aspects.

Table 14.1 Research work pertaining to big data in entire industry

Research	References
Advanced analytics-based petroleum asset management	[8]
Big data–based ship performance improvement study	[9]
Detection of invisible non-production time (INPT) using real-tile logging data	[10]
Modelling of energy efficiency and petroleum refinery estimation	[11]
Data Mining–based O&G risk assessment and online forewarning	[12]
Natural gas pipeline efficiency evaluation	[13]
PCA based air pollutant measurement in O&G operations	[14]
Reservoir modelling for unconventional oil and gas resources	[15,16]

Table 14.2 Real-world implementation of big data in entire industry

Companies	Implementation
Shell	Reduction in oil extraction costs by accurate understanding of what lies where underneath the surface; and ensuring lowest downtime and overhead costs
British Petroleum and Chevron	Locating of new reservoirs with high-resolution topographic maps of surface under the earth & beneath salt canopies
BioTechnologies and Glori Energy	Complementing of seismic data with novel sensors for economic benefit
HiFi Engineering and Silixa	Usage of hardware like connected tools & smart pumps for exploiting big data's potential for economic benefits
NuPhysicia	Providing telemedicine support to workers who are situated far from healthcare centres
Halliburton	Optimization of seismic and drilling space while attaining better well planning
Devon Energy	Determination and mitigation of cause for company's non-productive time in their operations
Infosys	Production and transportation services for O&G companies

14.11   SUPERIOR IMPLEMENTATION TRAITS

Keeping aforementioned real-world implementations in mind, there is an essential topic that needs to be discussed – what makes few companies better at big data implementation than others? There are indeed some aspects of the selected few that make their implementation "superior", which are summarized in Figure 14.8.

14.12   CHALLENGES

Until now, every aspect has portrayed a quite optimistic outlook to big data in the industry due to the benefits, diverse range of applications, extensive research work and several real-world implementations. Thus, the question that arises is – why is there not full-fledged big data implementation in the industry? The simple answer to that is the existence of problems, hurdles or challenges along the way that restricts or blocks the actual progress in big data's use in the industry by companies, which are summarized in Figure 14.9.

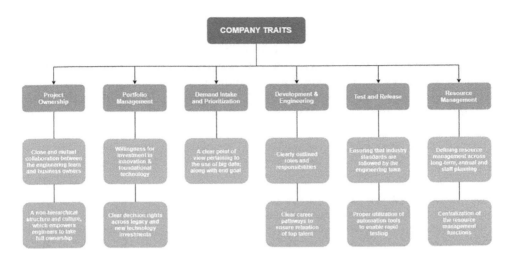

Figure 14.8 Traits of companies with successful big data implementation.

Figure 14.9 Major challenges of big data's implementation in petroleum operations.

14.13 FUTURE SCOPE

The previous section brought forward the existence of problems, hurdles or challenges that block the actual progress in big data's use in the industry by companies, after a complete course of optimistic outlook until then. Now that they are understood properly, it is time to dive into the future prospects of big data that pose as massive upsides to its real-world implementation, and if O&G companies undertake them effectively, they can reap the benefits that it offers in the long term. These aspects of future scope are summarized in Figure 14.10.

14.14 END

The book has clearly established the fact that big data plays an integral role now and will continue to do so in petroleum industry in the future. There has been significant development in the extent of applications, AI algorithms, platforms & tools, research

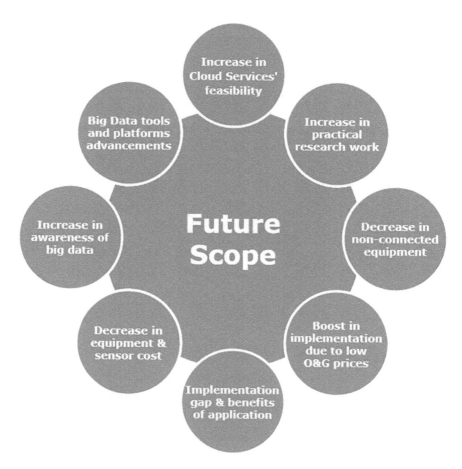

Figure 14.10 Future scope of big data's implementation in petroleum operations.

and real-world implementation of big data by O&G companies in the last few decades which paints an optimistic and viable (economically, environmentally and regulatory) viewpoint. The complete and fundamental integration of big data into petroleum industry is not an "IF" but a "WHEN"!

REFERENCES

1. Easley S. A. (2021). What Is Geoscience, Exactly? Southern New Hampshire University. https://www.snhu.edu/about-us/newsroom/2018/08/what-is-geoscience.
2. U.S. Geological Survey (2021). What Is Geoscience? Youth and Education in Science. https://www.usgs.gov/science-support/osqi/youth-education-science/what-geoscience.
3. Shah, N. K., Li, Z., & Ierapetritou, M. G. (2011). Petroleum refining operations: Key issues, advances, and opportunities. *Industrial & Engineering Chemistry Research*, 50(3), 1161–1170. https://doi.org/10.1021/ie1010004.
4. IBISWorld (2021). Global Biggest Industries by Revenue in 2021. https://www.ibisworld.com/global/industry-trends/biggest-industries-by-revenue/.
5. Kalyani, D. (2018). Oil Drilling & Gas Extraction in the US, IBISWorld Industry Report, 2018.
6. Schaafsma, S. Big Data: The 6 Vs You Need to Look at for Important Insights. Motivaction. https://www.motivaction.nl/en/news/blog/big-data-the-6-vs-you-need-to-look-at-for-important-insights.
7. Jain, A. (2016). The 5 V's of big data, Watson Health Perspectives, IBM. https://www.ibm.com/blogs/watson-health/the-5-vs-of-big-data/.
8. von Plate, M. (2016). Big Data Analytics for Prognostic Foresight. *SPE Intelligent Energy International Conference and Exhibition 2016*, 1–6. https://doi.org/10.2118/181037-ms.
9. Anagnostopoulos, A. (2018). Big Data Techniques for Ship Performance Study. *Proceedings of the 28th International Ocean and Polar Engineering Conference*, 887–893. https://onepetro.org/ISOPEIOPEC/proceedings-abstract/ISOPE18/All-ISOPE18/ISOPE-I-18-190/20389.
10. Yin, Q., Yang, J., Zhou, B., Jiang, M., Chen, X., Fu, C., Yan, L., Li, L., Li, Y., & Liu, Z. (2018). Improve the Drilling Operations Efficiency by the Big Data Mining of Real-Time Logging. *SPE/IADC Middle East Drilling Technology Conference and Exhibition 2018*, 1–12. https://doi.org/10.2118/189330-ms.
11. Azadeh, A., Ghaderi, S., & Asadzadeh, S. (2008). Energy efficiency modelling and estimation in petroleum refining industry – A comparison using physical data. *Renewable Energy and Power Quality Journal*, 1(6), 123–128. https://doi.org/10.24084/repqj06.242.
12. Wang, T., Li, T., Xia, Y., Zhang, Z., & Jin, S. (2017). Risk assessment and online forewarning of oil & gas storage and transportation facilities based on data mining. *Procedia Computer Science*, 112, 1945–1953. https://doi.org/10.1016/j.procs.2017.08.052.
13. Fan, M. W., Ao, C. C., & Wang, X. R. (2019). Comprehensive method of natural gas pipeline efficiency evaluation based on energy and big data analysis. *Energy*, 188, 1–12. https://doi.org/10.1016/j.energy.2019.116069.
14. Tokarek, T. W., Odame-Ankrah, C. A., Huo, J. A., McLaren, R., Lee, A. K. Y., Adam, M. G., Willis, M. D., Abbatt, J. P. D., Mihele, C., Darlington, A., Mittermeier, R. L., Strawbridge, K., Hayden, K. L., Olfert, J. S., Schnitzler, E. G., Brownsey, D. K., Assad, F. V., Wentworth, G. R., Tevlin, A. G., . . . Osthoff, H. D. (2018). Principal component analysis of summertime ground site measurements in the Athabasca oil sands with a focus on analytically unresolved intermediate-volatility organic compounds. *Atmospheric Chemistry and Physics*, 18(24), 17819–17841. https://doi.org/10.5194/acp-18-17819-2018.

15. Lin, A. (2014). Principles of Big Data Algorithms and Application for Unconventional Oil and Gas Resources. *SPE Large Scale Computing and Big Data Challenges in Reservoir Simulation Conference and Exhibition 2014*, 1–9. https://doi.org/10.2118/172982-ms.
16. Chelmis, C., Zhao, J., Sorathia, V., Agarwal, S., & Prasanna, V. (2012). Semiautomatic, Semantic Assistance to Manual Curation of Data in Smart Oil Fields. *SPE Western Region Meeting 2012*, 1–18. https://doi.org/10.2118/153271-ms.

Index